Practical Accounting & Financial Management

for Farm and Small Business

Ben Brown

The Crowood Press

Practical Accounting
& Financial Management

for Farm and Small Business

First published in 1991 by Farming Press Books, Ipswich

This edition published in 2003 by
The Crowood Press Ltd
Ramsbury, Marlborough
Wiltshire SN8 2HR

www.crowood.com

British Library Cataloguing-in-Publication Data
A catalogue record for this book is available from the British Library.

ISBN 1 86126 641 3

Acknowledgements
Grateful thanks to Anne Johnson, Chris Ward, Paul Green and Don and Jackie Griffin, who checked the figures, read script and generally gave encouragement; and to the Peacock Inn, Redmile, Notts.

Typeset by Cambridge Photosetting Services

Printed and bound in Great Britain by CPI Bath

Contents

Preface

I set out to produce a book that is readable with basic concepts clearly explained. The accounting and financial management techniques covered can be applied to almost any business, be it farming, manufacture, building, services or wholesale and retail outlets. As I often reminded my students, the skills they gained were applicable to any business activity, from farming to fashion.

The case studies are based on farming and a garden centre. However, it is the financial aspects that we are considering. Bricks or books can replace animal feed or herbicides, garments or bicycles can replace milk or potates, with regard to physical description of units of cost and output. Finance is the common theme for them all.

No attempt has been made to include taxation, either value added tax or income tax, other than through brief reference since they are matters which are subject to constant change and revision. Customs and Excise and the Board of Inland Revenue publish details of regulations and guides on implementation and it is from these and the specialist consultant that the reader should seek further information and advice.

The VAT rate changes from time to time and this occurred just as the book was going to press. Accounting method is not affected by the VAT rate; therefore the relevance of the book is unaltered by changes which have already occurred or which may occur in the future.

Administration of VAT is subject to many regulations and failure to comply may lead to heavy financial penalties. The reader is therefore directed to current literature produced by Customs and Excise (VAT) for guidance.

This book is suitable for the complete beginner with no previous knowledge of accounting but the reader moves quickly and easily on to sophisticated techniques that are explained in the simplest terms. It is a fact that the majority of business failures are due to cash-flow problems; financial management, including production of the cash-flow budget, is the underlying theme of the book.

Ben Brown
March 2003

CHAPTER 1

What is it all about?

Book-keeping, accounts, accounting, balance sheets: what pictures do these raise in your mind? Is it something very hazy, columns of meaningless typed figures and obscure titles? Relax, you are not alone; even some business proprietors suffer this problem quite unnecessarily. The following chapters consider each stage in the accounting procedure in a basic logical way, but first we take a brief look at the complete picture in order to gain a sense of direction.

The balance sheet

From time to time it is necessary to effectively stop the business for a brief spell and determine exactly where it stands financially, what it owes and what it owns. This calculation is known as a balance sheet, or more descriptively a statement of affairs. Clearly, this information will not only be a requirement at the commencement of business but also at the end of any period of trading. It is in fact the most important and telling document produced for any business. Remember, though, that it gives a picture of that business at one point in time.

Books and book-keeping

A starting point for business is established with the balance sheet, but what happens from there until the end of the year or the trading period? In simple terms, a day to day record of trading is built up from payments and receipts or purchases and sales. At the end of the trading period a summary of purchases and sales is made and this provides a picture of transactions during the year. Book-keeping is financial data collection over a set period of time, but further calculations are necessary before we reach the next balance sheet.

Year end accounts – the profit and loss account

Book-keeping provides detail of transactions for the year, purchase of inputs and sale of outputs, but this is not sufficient to show the result of business activities in terms of profit or loss, since other factors are

involved as well. Profit may be hidden in increased value of stock and vice versa. Estimated fall in value of fixed assets such as machinery must be taken into account as a cost and the value of products taken for the personal use of owner or workforce must be included as part of output. The profit and loss account takes all of these factors into consideration in calculating profit or loss for the trading period, the results being taken to the capital account which is drawn up at the same time as the new balance sheet at the year end.

The capital account

The capital account shows the factors which contribute to increase or decrease of capital from the start of the year, taken from the opening balance sheet, to the end of the year. Included in the calculation are opening capital, to which is added profit and introduction of funds; loss and withdrawal of funds are deducted to show the new capital which is also shown in the balance sheet for the year end.

The year end or closing balance sheet

A new balance sheet is produced at the end of the trading period and this completes the cycle of basic accounts. The capital account provides the link between opening and closing balance sheets. The closing balance sheet produced at the year end is also the opening balance sheet for the following trading period.

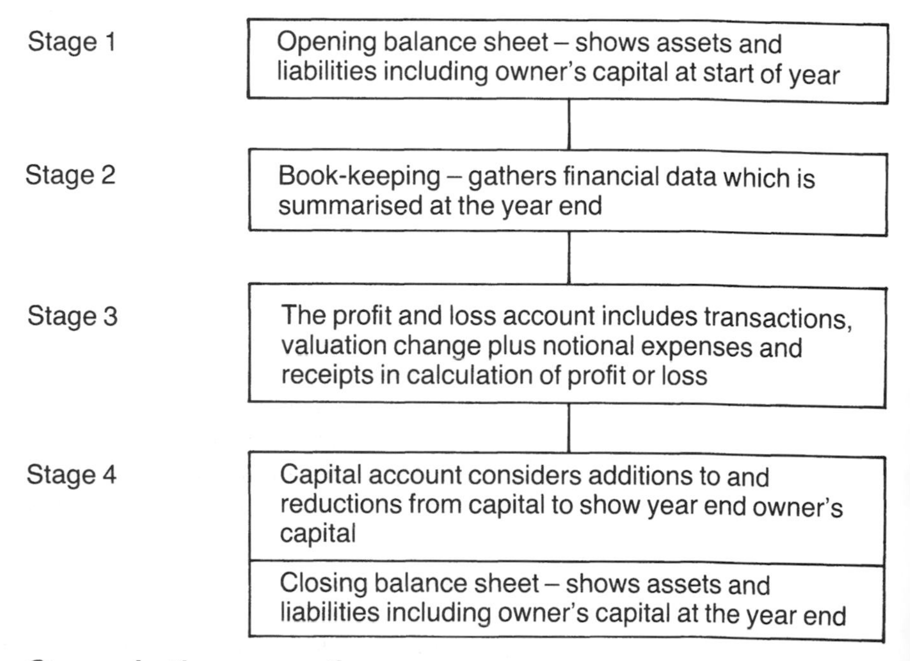

Stages in the accounting year

Where next?

There is more, much more, but all the time there is a logical sequence of events and a clear and distinct purpose for each procedure.

Flow of funds statement

The profit and loss account shows profit or loss made during the year and the balance sheet shows owner's capital, but the state of the bank account often does not appear to reflect the result of trading for the year. The flow of funds statement shows what has happened to **money**, whether frozen or liquified in a number of ways. It shows where cash in the bank has come from and where it has gone. This could suitably be shown as Stage 5 in the flow chart above.

How well have we done? – Analysis of the accounts

The profit and loss account, although showing overall performance in terms of profit or loss, is an obscure document. Satisfactory levels of achievement in one part of the business may be masked by failure in other areas and so the accounts must be analysed in order to determine exactly what is going on.

Gross margin analysis

Gross margin analysis achieves two things in particular: it separates overhead costs from costs which are directly related to production and it separates performance of individual enterprises so that each one can be seen and judged accordingly. Margins and fixed costs can then be used to produce a gross margin account which shows the same profit as the profit and loss account, but in a much more informative way.

Plans and budgets

Planning is nothing new – everybody does it but in varying degrees of formality and with varying degrees of success; but formal planning becomes increasingly important as competition in the market place increases. Plans and budgets are based on known past performance and adjusted to take changes into consideration. Planning and budgeting in many ways can be seen as setting out the direction in which it is intended to travel, with a series of marker posts.

Monitoring and control

Where have we been? Have we stuck to our plan and budget? If not, why not? What can we do to correct the situation? All these are questions to be asked and answered in ensuring that the business is kept on track.

The overall view

The broad picture has now been presented and will be developed stage by stage in the following chapters, starting with consideration of the balance sheet. Initial examples are simple in order to concentrate on concept and method rather than risk getting lost in detail.

CHAPTER 2

The balance sheet

What is he worth? This is an age old question, no doubt, but one which we all ask at times owing to universal curiosity regarding the material possessions of our fellow beings. The document which reveals all – and hence is the most closely guarded – is the balance sheet. In simple terms it is the record of everything which is **owned** by the business and everything which is **owed** by the business. The uninitiated often find it rather confusing to see that the total amount owned – the assets – is equal to the total amount **owed** – the liabilities – but there is an explanation for this.

Imagine you own a small business. You make a list of all the business assets, the equipment, goods, money and debts owed to you and calculate the total value. You then list and total your liabilities, the debts payable to your creditors and loans repayable by you. Subtract the total liabilities from the total assets and the resulting amount is what is left for you. This is the balance which represents capital or equity – the owner's share – and is shown with the liabilities, so making total liabilities equal in amount to total assets.

The owner and the business

A basic concept in accounting always separates the owner(s) from the business. The owner, or more correctly the proprietor, invests capital in the business and the business therefore has a liability to the proprietor. The fact that the proprietor contributes skill and effort to the business, and may in fact be the only person working in the business, makes no difference. This is known as the **business equity** concept.

The balance sheet – one point in time

The record of assets and liabilities which make up a balance sheet is specific to one point in time, not a period of time. A balance sheet can be considered as a sort of "stopping off" point which marks the beginning or end of a trading period. Each balance sheet is a record or picture of the financial state of the business on the date given at the head of the document. For example:

Balance Sheet for John Bull at 6th April 199X

Assets	£	*Liabilities*	£
Equipment	1,000	Capital	1,050
Stores	500		
Bank balance	200	Bank loan	400
Debtors	50	Creditors	300
	1,750		1,750

So we see that the total amount owned – the **assets** – is equal to the total amount owed – the **liabilities** – including that which is 'owed' to the proprietor. The size of the owner's share in the business is usually a closely guarded secret unless it is a public limited company.

The format of this balance sheet is just one simple example. The two lists may change sides or the asset list may be positioned above the liability list, but the information remains the same. If the content is understood, differing format does not create a problem.

Coming to terms with terminology

Already the necessity to use special terms has arisen. As with any specialist area of experience, be it music, medicine, sailing or others, the need to communicate clearly and specifically gives rise to a 'jargon' or special terminology. Here we consider some of the terms used relative to the balance sheet and business in general.

ASSETS

Broadly speaking, the assets of a business are all those things belonging to the business which are of value. These include the obvious, such as land, buildings, equipment, stock in hand and money, together with money owed to the business by debtors.

Classification of assets

From the brief list of assets given above it can be seen that they vary considerably in description. Land can be considered permanent, buildings are normally expected to have a long life, equipment usually wears out after a few years but stock in hand, which may include feed and fertiliser on a farm or sale items in a shop, are 'here today and gone tomorrow'. There is a difference in their level of permanence and therefore their function in the business. Assets are classified according to their function and their availability for conversion into cash.

Fixed assets

An ice-cream man has a very simple business consisting of a van fitted with refrigerators from which he sells ice cream. His fixed assets are the van and equipment; if he disposes of them the business cannot function. Likewise, a farm business cannot continue to operate without land, buildings, machinery and the cows that produce calves and milk, sows that produce piglets and ewes that produce lambs.

Fixed assets are then those assets which enable the productive process to be carried out. As the knitting machine is to the woollens' manufacturer, so is the cow to the dairy farmer. No knitting machine – no jumpers; no cow – no milk. The reason for the term fixed is becoming clear, since it can be seen that without the fixed assets the business cannot function. No golden eggs without the goose!

Current assets

If we take a further look at the ice-cream business, we find that the current assets are stocks of ice cream, wafers, cornets and wrappers which the proprietor hopes to sell, together with money in the till and bank. If he has supplied ice cream for which he has not been paid, then these debts are current assets also. Likewise, for other business examples, anything which is produced or held for sale such as grain, meat, milk, vegetables and wool is a current asset. Raw materials used in the productive process, for example feed, fertiliser, insecticides and even tractor fuel, are also current assets since they are intended for sale when they have become part of, or contributed to, the end sale product. The important point is that they are all in the process of being turned back into cash.

Liquid assets

Money held 'in hand' or in the bank, together with money owed to the business, represents the current assets which have 'liquified' through being sold. The money and debts owed to the business are current assets but are sometimes also sub-classified as liquid assets for reasons to be considered later.

LIABILITIES

If the ice-cream vendor owes money for stocks of ice cream, he will be required to pay up fairly quickly – he has a short term or current liability. A loan for purchase of his van, repayable over several years, would be a long term liability.

Long term liabilities

Any mortgage, loan or purchase which is not due for settlement within the next accounting year can be classified as long term. Extended credit on a purchase is usually arranged as a loan. Such liabilities will normally have been incurred to enable purchase or improvement of fixed assets.

Classification of assets and liabilities: example

Balance Sheet for a Growing Business as at 1.1.199X

Fixed assets	£	*Long term liabilities*	£
Machinery		Capital	
Fixed equipment		Mortgage	
Vehicles		Bank loan	
Dairy cows		HP loan	
Current assets		*Current liabilities*	
Young livestock		Sundry creditors	
Beef cattle			
Wheat			
Feedstuffs			
Fertiliser		* Bank (–bal.)	
Sprays			
Sundry stores			
Bank (+bal.)*			
Cash in hand			
Sundry debtors			
	______		______
	======		======

* The bank position will appear either as an asset (+ bal.) or as a liability (– bal., overdraft)

Current liabilities

These are liabilities which are incurred in the day to day process of trading and are due to be settled within the next accounting year. Generally, the list will include purchased but not yet paid for items of stock of raw materials, goods for re-sale and services, e.g. electricity and rates. The total amount owed to creditors is summarised under a heading of sundry creditors for inclusion in the balance sheet; it is pointless and usually impractical to include a list of creditors.

A bank overdraft is also classified as a current liability since the facility is always subject to withdrawal at short notice.

FORMAT AND STYLE OF BALANCE SHEETS

Balance sheets may be produced in a variety of formats, with assets on the left or on the right side of the page. Historically, there was a statutory requirement to show assets on the right in the UK, but this is no longer the case. Assets still commonly appear on the right in the

UK but there is a gradual trend towards the international convention which puts assets on the left. The vertical or narrative style which puts assets at the top and liabilities below is perhaps more logical than either of the above and is becoming increasingly common.

Take a look at the published accounts of public limited companies in the national press. You will find a wide variety of styles. The important issue for the reader is the ability to recognise the *content* of the different sections, whether on left or right, above or below.

In order to move with the times, the international convention is followed in this book. A number of examples of balance sheet formats follow.

Traditional British
Balance Sheet for .... at ....

	£		£
Capital		F. assets	
L.T. liabilities			
		C. assets	
S.T. liabilities			
	____		____
	====		====

International convention
Balance Sheet for .... at ....

	£		£
F. assets		Capital	
		L.T. liabilities	
C. assets			
	____	S.T. liabilities	____
	====		====

Narrative or vertical – simple
Balance Sheet for .... at ....

	£
F. assets	
C. assets	____
Total assets	____
L.T. liabilities	
C. liabilities	____
Total liabilities	____
Capital (=Assets–Liabilities)	====

More developed
Balance Sheet for .... at ....

	£
F. assets	____
C. assets	____
Less C. liabilities	____
Net C. assets	____
(less)	____
L.T. liabilities	====
Capital (detail)	====

It is also normal to include profit (or loss) from the past year of trading in order to show progress from the previous balance sheet. I have deliberately omitted this complication so that the reader concentrates on balance sheet content only at this stage.

WHAT WILL A BALANCE SHEET TELL US?

Firstly, it may be helpful to introduce alternative descriptions of this document. A balance sheet may also be called a *position statement* or a *statement of affairs*. The important point is that the information contained in the balance sheet refers to the position at one stated point in time. A balance sheet is always dated to refer to one day only, not a period of time.

One balance sheet in isolation will tell us something of the present strengths and weaknesses of the business. The proportion of (own) capital relative to borrowed capital is an important measure. The ability to pay current liabilities with current assets is also of prime importance.

A series of balance sheets produced over several year ends will show in which direction the business is moving: up or down. Strengths and weaknesses can be compared between balance sheets to produce a trend which may be good news or may provide early warning of necessity for change. More on balance sheet analysis at a later stage.

BALANCE SHEET TRANSACTIONS

Balance sheets are normally only drawn up at the end of a trading period, but here we consider the impact of individual transactions on balance sheet detail. For simplicity of example, we will consider the part time business of I. Cornetto, the ice-cream vendor referred to earlier. The balance sheet for his business at 1 June 199X appears below:

Balance Sheet for I. Cornetto at 1.6.199X

	£	£		£
Fixed assets			Capital	2,360
Van	3,000			
Refrigerator	800	3,800	*L.T. liabilities*	
			Bank loan	2,000
Current assets				
Ice cream	200			
Wafers etc.	20		*Current liabilities*	
Sundry	10	230	Sundry creditors	120
Bank	400			
Cash in hand	50	450		
		4,480		4,480

On 3 June Mr Cornetto paid the £120 owed to creditors from his bank account. This has the effect of reducing the bank balance by

£120 and the current liabilities are reduced by the same amount. The two entries balance each other and capital remains the same.

Balance Sheet for I. Cornetto at 3.6.199X

	£	£		£	£
Fixed assets			Capital	2,360	
Van	3,000				
Refrigerator	800	3,800	*L.T. liabilities*		
			Bank loan		2,000
Current assets					
Ice cream	200		*Current liabilities*		
Wafers etc.	20		Sundry creditors		–
Sundry	10	230			
Bank	280				
Cash in hand	50	330			
		4,360			4,360

On 4 June Mr Cornetto purchased *on credit* an additional £50 worth of ice cream in anticipation of a busy weekend.

Balance Sheet for I. Cornetto at 4.6.199X

	£	£		£
Fixed assets			Capital	2,360
Van	3,000			
Refrigerator	800	3,800	*L.T. liabilities*	
			Bank loan	2,000
Current assets				
Ice cream	250			
Wafers etc.	20		*S.T. liabilities*	
Sundry	10	280	Sundry creditors	50
Bank	280			
Cash in hand	50	330		
		4,410		4,410

The purchased stock is added to current assets at cost price and the debt incurred is entered under current liabilities. Again, the two entries balance out and so capital is unchanged.

On 5 June Mr Cornetto has a good day and sells £60 (cost) worth of ice cream together with £5 (cost) worth of wafers and cornets at a mark up of 100 per cent on cost, i.e. double the cost price. Total sales of £130 were achieved. A balance sheet at the end of the day would show:

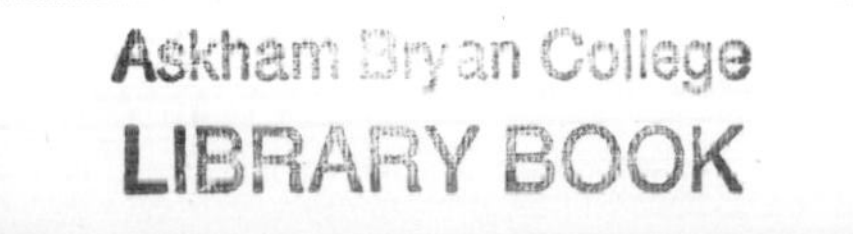

Balance Sheet for I. Cornetto at 5.6.199X

	£	£		£
Fixed assets			Capital	2,425
Van	3,000			
Refrigerator	500	3,800	*L.T. liabilities*	
			Bank loan	2,000
Current assets				
Ice cream	190		*Current liabilities*	
Wafers etc.	15		Sundry creditors	50
Sundry	10	215		
Bank	280			
Cash in hand	180	460		
		4,475		4,475

Cash in hand has increased by £130, stock is reduced by £65. The £65 profit margin made on sales (sales £130 minus cost of sales £65 = £65 profit margin) has the effect of increasing equity by that amount.

If we now move forward to the end of the summer and produce a balance sheet for 30 September, the profit made over the three months can be calculated from change in capital. Note: In this simple example there is no depreciation (fall in value) of the fixed assets. No cash is drawn from the business for private (non-business) use.

Balance Sheet for I. Cornetto at 30.9.199X

	£	£		£	£
Fixed assets			Capital at 1.6.199X	2,360	
Van	3,000		Profit for period	500	
Refrigerator	800	3,800	Capital		2,860
Current assets			*L.T. liabilities*		
Ice cream	50		Bank loan		2,000
Wafers etc.	10				
Sundry	5	65	*Current liabilities*		
Bank	1,025		Sundry creditors		80
Cash in hand	50	1,075			
		4,940			4,940

Capital has increased through selling stock at a profit of £500.

Balance sheet transactions – summary

Profit is not normally calculated in the method demonstrated for Mr Cornetto since there is a very unsatisfactory lack of detailed information – the tax inspector would be quick to agree with this! But the series of simple exercises gives at least an insight into the mechanism of balance sheets.

Balance sheets are normally drawn up at the end of a period of trading, therefore once annually. Interim accounts and balance sheets may be produced half-yearly or quarterly according to the requirement for good business management. Balance sheets may also be drawn up on other occasions from time to time, either on request from a bank manager as a form of progress report or perhaps in support of an application for a loan.

Progress to the next chapter

In the examples considered in this chapter we looked briefly at the day to day progress of the ice-cream vendor through a series of balance sheets. Clearly it would be tedious and impractical to maintain financial records in this manner. How then do we account for the period of time between balance sheets, the period of trading? In the following chapter we consider construction of the trading and profit and loss account together with the records which make this possible.

If a confident understanding of each stage of work is to be gained, the reader will find it helpful to complete the exercises given and develop them beyond the information provided with 'what if?' questions – there is limitless scope here for a fertile imagination! Make further adjustments to Mr Cornetto's balance sheets – pay off creditors, take out a loan, buy a freezer, sell stock – use your imagination and see the effect it has on the balance sheet.

Draw up a personal balance sheet to include your material possessions which have a realistic money value, together with your financial liabilities. Do not forget that you have produced a highly confidential document!

CHAPTER 3

The profit and loss account

In this chapter we look at what goes on in between balance sheets during the period of trading, which is usually one year. Success or failure of the year of activity is summarised in the profit and loss account and is indicated as either profit or loss. Calculation of profit is not simply a question of adding up payments and deducting that total from total receipts, there are a number of factors which must be considered and we take them stage by stage below.

PROFIT – OR LOSS

The main factors which must be considered in order to determine profit level are:

valuation change
revenue and expenditure
notional expenses

Each of the three factors is built into a normal profit and loss account for businesses large and small across a range of activities from farming to fashion – it is all business, and business performance is stated in financial terms.

Valuation change

Profit does not necessarily lead to more money in the bank and, in the short term, loss does not necessarily deplete the bank account. We consider why.

Productive activity will normally add value to a commodity, whether it be feeding and caring for a growing animal or the construction of something, for example a gate. These products are not necessarily ready for sale before the end of a trading period and so the 'added value' must be estimated for inclusion in the accounts. A word of caution here though; valuation of assets must be carried out carefully and consistently, following established guidelines. Over-estimation of value will lead to disappointment later when the item is sold at a price lower than hoped for.

A simple example of valuation increase is shown below in the case of a growing animal. A beef calf is purchased at a few days old for £100. It is fed and cared for over a period of twelve months and is then valued at £250, showing an increase in value of £150. When the cost of food and sundries is deducted, a 'profit' of £60 remains – but it is still walking about as part of the beast.

	£
Value of 1 yr beast	250
Less cost of calf	100
Increase in value over 12 months	150
Less cost of food and sundries	90
'Profit'	60

Clearly, the farmer's wife can't buy groceries with 'profit' that is still part of a live animal. The same is true of the manufacturer who has partially completed and finished goods in store, unsold at the end of a trading period. The products are certainly worth more than a heap of new materials due to the input of skill and effort – but the profit is tied up until the goods are sold.

Revenue and expenditure

Suffice it to say at this stage that 'revenue' represents true receipts and 'expenditure' is the true amount of expense for the trading period.

Sale of produce, skill and effort result in receipt of money and it is these essential business activities which generate profit in cash after deduction of expense. This, then, is the second factor which we have to consider when attempting to calculate profit.

The example of a market trader selling vegetables which he has purchased from a wholesaler will illustrate the basic principles of revenue and expenditure. The trader may purchase stock in the morning and sell it all during the day. From the total value of sales must be deducted the 'cost of sales'. In this case it is the amount he paid for the vegetables that morning. In addition, any payments he has to make for using the market stall plus sundry costs such as wrapping paper and transport must all be deducted from the sales figure in order to determine profit.

Profit and loss account for Mr Green (for one day of trading) 4 June 19X8

Expenditure	£	*Revenue*	£
Purchase of vegetables	100	Sale of vegetables	200
Hire of stall	20		
Sundry costs	5		
Transport	5		
Total expenditure	130		
Profit	70		
	200		200

Another way of presenting that account would be as follows – but whether expenditure and revenue are side by side or one above the other, it makes no difference to the calculation.

Revenue	£	£
Sale of vegetables	200	
Expenditure		
Purchase of vegetables	100	
Hire of stall	20	
Sundry costs	5	
Transport	5	130
Profit		70

Calculation of profit becomes a little more complex when stocks of goods and equipment are carried forward from one trading period to the next. A value must be fixed for all of the items and this effectively becomes the initial investment for the new trading period. Likewise, at the end of the trading period, any remaining stocks held are valued and entered in the profit and loss account as the final transaction for the trading period. Equipment is entered in the closing balance sheet.

Mr Green gets a bit more adventurous and buys more vegetables than he can sell in a day.

Profit and loss account for (one day of trading) 5 June 19X8

Expenditure	£	*Revenue*	£
Purchase of vegetables	150	Sale of vegetables	220
Hire of stall	20		
Sundry costs	5	*Closing valuation*	
Transport	5	Stock of vegetables	40
	180		
Profit	80		
	260		260

He has made a profit from the day's trading, but he still has £40 tied up in stock at the end of the day leaving only £40 'surplus cash'. If the remaining stock was unsold for any reason, the full profit of £80 would never be realised. The following day:

Profit and loss account for (one day of trading) 6 June 19X8

	£		£
Opening valuation		*Revenue*	
Stock of vegetables	40	Sale of vegetables	240
Expenditure			
Purchase of vegetables	130	*Closing valuation*	
Hire of stall	20	Stock of vegetables	50
Sundry costs	10		
Transport	5		
	205		
Profit	85		
	290		290

He has now reached a situation where stock is brought forward at valuation from the previous day (or trading period) and where stock remaining at the end of the day is valued and entered in the account before being carried forward to the next day (or trading period). Closing stock is, in effect, 'sold' to the following trading period.

Note: The day by day calculation of profit in this way is not realistic or sensible, but the same principles apply whether the trading period be days, months or a year.

Valuation has entered our calculation of profit but in this example goods are simply purchased for resale, nothing is done to them to make them of greater value before being sold and so they are valued at cost price for entry in the profit and loss account. Since the goods do not grow in value before being sold, there is no element of profit tied up in the closing valuation other than investment in goods at cost price.

Another way of presenting the accounts is as follows and here we divide our calculations into two stages but the end result is the same:

Trading profit and loss account for (one day of trading) 6 June 19X8

	£		£
Opening stock of vegetables	40	Sale of vegetables	240
Purchase of vegetables	130		
Total available for resale	170		
Closing stock of vegetables	50		
Cost of vegetables sold	120		
Gross profit c.f.	120		
	240		240

Sundry expenditure		Gross profit	120
Hire of stall	20		
Sundry costs	10		
Transport	5		
Total expenditure	35		
Net profit	85		
	120		120

Students of accounting often find it difficult to understand why profit appears under expenditure and why by some mysterious process we finish up with identical figures on both sides of the most conventional 'side by side' presentation. Really, there is no mystery. Convention has it that we must 'balance' our accounts, so the balancing figure represents the difference between revenue and expenditure – profit. Therefore the sum of profit and expenditure is identical to revenue.

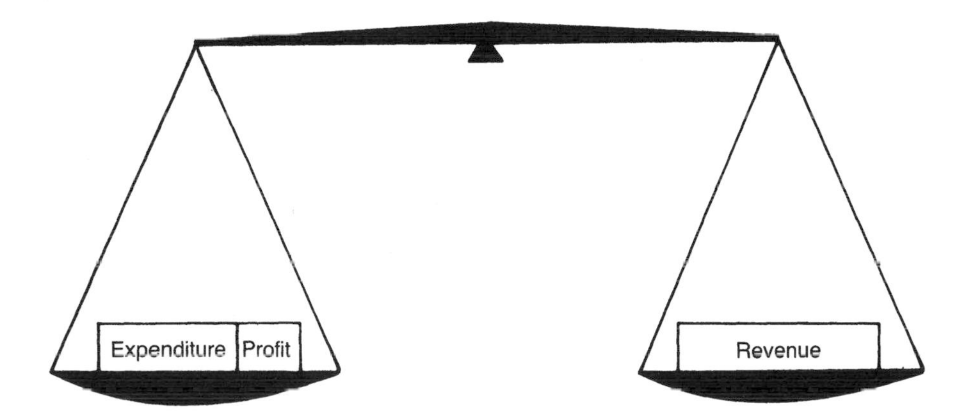

Figure 3.1 'Balancing' the accounts

Let us have another look at the budding farmer who raised the beef animal. During his second year of trading he sells his finished animal for slaughter. He also buys more young stock for rearing and finishing. These cattle are intensively reared in buildings for which he pays rent.

Profit and loss account for B. Farmer for year ending 5th April 19X9

Opening valuation	£	*Revenue*	£
1 one-year-old beast	250	1 fat beast	400
Expenditure			
Calves (5)	500		
Feeding stuffs	300		
Sundry livestock costs	65	*Closing valuation*	
Rent	50	5 cattle	1,250
Transport	50		
(val. plus total expend.)	1,215		
Profit	435		
	1,650		1,650

We now have a situation where 'value' tied up in a beast at the start of the year is turned into cash when it is sold. More young cattle were purchased and fed and it is easy to see that the investment in the business is far greater at the end of the year than at the beginning – all of the profit made on the beast which was sold has been re-invested – no wild spending spree for this young farmer!

Notional expenses – depreciation

The part-time farmer for whom we have just produced an account runs an old pick-up truck which serves as both private transport and for business purposes. This truck, like all vehicles, deteriorates through 'wear, tear and age' and so falls in value or 'depreciates'. The fall in value cannot be established exactly unless the vehicle is sold and so the fall, or depreciation must be estimated. Vehicles and equipment are not entered into the profit and loss account when they are purchased, since they normally have a useful life which spans several trading periods. It is only the **estimated fall in value** which is entered into the account as a cost, since no payment has been made by the business which represents the fall in value over one year. This cost then is known as a 'notional cost' – it is only a notion or 'guesstimate' rather than a specific payment, but it must be taken into consideration and will affect the profit for the year.

If we give the pick-up truck a value of £1,000 and a life of 5 years, after which it will be scrapped, there will be an annual fall in value or 'depreciation' of £200. The young farmer reckons that about 25 per cent of his mileage is for his enterprise and so it is reasonable to charge the business with 25 per cent of the annual depreciation. £200 × 25% = £50. We show this as an expense in the revised account below.

Profit and loss account for B. Farmer for year ending 5th April 19X9

Opening valuation	£	*Revenue*	£
1 one-year-old beast	250	1 fat beast	400
Expenditure			
Calves (5)	500		
Feeding stuffs	300		
Sundry livestock costs	65	*Closing valuation*	
Rent	50	5 cattle	1,250
Transport	50		
	1,215		
Depreciation on truck	50		
	1,265		
Profit	385		
	1,650		1,650

THE CAPITAL ACCOUNT

We have considered two major stages in the accounting process, the balance sheet and then the profit and loss account. Now we must link together the progress through the balance sheet at the start of the year, the profit and loss account which shows the performance during the year and finally a balance sheet at the end of the year.

The part-time farmer for whom the profit and loss account above was produced settles all of his debts at the time of purchase or sale and so has no debtors or creditors. He started his second year of trading with £1,000 in the bank and as we have already discovered, he has a pick-up truck which was worth £1,000 at the beginning of the year.

Balance sheet for B. Farmer at (start of year) 6th April 19X8

Assets	£	*Liabilities*	£
1 beast	250	Capital	2,250
1 pick-up truck	1,000	Loans	–
Bank balance	1,000	Sundry creditors	–
	2,250		2,250

Profit for the year, taken from the profit and loss account, is £385.

Balance sheet for B. Farmer at (end of year) 5th April 19X9

Assets	£	*Liabilities*	£
5 cattle	1,250	Capital	2,485
1 pick-up truck	800	Loans	–
Bank balance (see summary of bank transactions for year below)	435	Sundry creditors	–
	2,485		2,485

Bank	£
Bank balance at start of the year	1,000
Receipts for the year	400
	1,400
Less	
Payments for the year	965
Bank balance at the end of the year	435

In the capital account which follows we see the progression from start to end of the year. In this case we had a situation where 75 per cent of the fall in value of the truck was regarded as being due to private use (this would not happen other than in such a part-time business) and so we shall consider that 'value' has been taken from the business by private wear and tear on the truck. Total depreciation was £200, £50 was allowed as a business notional expense and so the remaining £150 fall in value will be regarded as private drawing since it was due to private use of the vehicle. There are alternative ways in which the truck could have been included but for the time being we will consider the effect of this approach.

Capital account for year ending 5th April 19X9

	£
Capital at start	2,250
Profit for the year	385
	2,635
Less	
'Drawings' (private part of truck depreciation)	150
Capital at end of year	2,485

The capital account shows what has happened to capital during the year, how it has been added to and how it has been depleted. The layman often refers to it as a proof of accounts, but this ignores the true purpose of this document.

CHAPTER 4

Book-keeping

Book-keeping is the systematic recording of business transactions: buying and selling, payment and receipt. These activities normally occur continuously throughout the business year and so detail must be collected in an orderly way which allows accurate totals to be determined at the year end. The books so produced bridge the gap between business year ends by collecting financial data from which both profit and loss account and balance sheet can be drawn up.

The method of book-keeping used will be determined by the type of business. In cases where goods and services are supplied on the basis of immediate settlement or where only a relatively small number of customers and suppliers are involved, then a **cash based system** is sufficient. In cases where goods and services are supplied to, and purchased from, a large number of firms and individuals on credit, it becomes necessary to adopt a **double entry system**.

Meaning of 'cash'

It is necessary to consider for a moment the meaning of 'cash' relative to book-keeping. Settlement of debt by cheque, credit transfers, credit card, bank notes and coinage are all considered to be cash for book-keeping purposes since they all represent a flow of money from one person or business to another person or business.

CASH BASED SYSTEMS

Simple cash book

For the very simple business, where there is no requirement to break down or analyse payments and receipts, all that is needed is a single column cash book into which is entered basic detail of the transaction and the amount of money. A small retail business such as a village shop or corner shop may simply enter the total amount of takings for the day or week under receipts. Separation of payments for overheads such as rent, from payments for raw materials or goods for resale may well be done at the year end. The important point is that all payments and all receipts should be recorded in the order in which they occur and all invoices must be filed and preferably coded in the same sequence as the record.

This method must be regarded as very basic. It does ensure that affairs are kept in tidy form and it also enables a check on the bank funds available. The advantage of this simple record is that it is quick to maintain, but it has little use other than for calculation of cash flow and a crude profit and loss account. More detail is required in order that tax and informative management accounts may be produced.

CASH ANALYSIS

The simple cash book described above provides only a total of payments and receipts with no regard to different categories into which they fall. Payments for a variety of purposes such as rent, motor expenses, purchase of stock and administration are lost in one sum total and so it becomes difficult to identify strengths and weaknesses of the business. Excessive total costs could be due to overspending on just one or more of these items, but how to determine the culprit?

Cash analysis is an extension of the simple cash book. The 'Bank' column records total payments and total receipts, which allows a check to be made against the bank statement – the bank reconciliation. In addition to this, every item is recorded again in one or more separate columns farther across the page but on the same line. By this means the payments and receipts are analysed or broken down into different categories according to need, which then allows detailed accounts to be produced.

Column headings

The first task when creating a new cash analysis record is to decide how payments and receipts are to be analysed, or in other words what column headings or names will be used. This will depend on the type of business and what information is required from the accounts but the issue will be further considered at a later stage.

Use of the bank and the 'Bank' column in the analysis book

All business payments and receipts should be channelled through a bank account. If cash is received from sales, it should be paid into the bank account rather than used to pay bills. Payments should be made by cheque, credit transfer, direct debit or banker's order.

When payments and receipts pass through a bank account in this way, the bank statement becomes a valuable record in its own right and provides a means of checking the accuracy of accounting books.

The bank statement is sometimes used as the main record of payments and receipts but there is a disadvantage: cheques written and sent to various individuals and firms will be paid into their bank accounts and eventually the amounts will be deducted from the business account, but the order in which they appear on the bank statement will vary considerably from the order in which cheques were written. There may be a delay of weeks or even months before amounts are deducted from the account. It should be noted that if a cheque is six months old or older it is regarded as 'dead' and banks will not accept it.

Because of these delays, the balance shown on a bank statement does not normally indicate the true, up-to-date position of the account. The only way in which an up-to-date record of payments, receipts and bank funds available can be maintained is by recording all cheques paid out, at the time and in the sequence in which they were written. Similarly cheques and money paid into the bank should be recorded in sequence. A list of credit transfers, direct debits and bankers' orders due should be available which will allow these payments and receipts to be entered in the record as well to give a complete and accurate picture.

Complicated? Well, if taken a step at a time it is quite simple but basic rules must be followed:

1. Record cheque payments in cheque number sequence.
2. Record credit transfers, direct debits and bankers' orders due for the period (week, month or quarter dependent on recording method).
3. Record all payments into the bank in paying-in book sequence.
4. Record payments into the bank made by credit transfer. When the amount varies, it will be necessary to await written notification, e.g. for the M.M.B. milk cheque, although it may be possible to determine the amount via a computer link or by telephone.

'Bank' column in a cash book

As explained earlier, all payments whether from a bank account or by bank notes and coinage are regarded as 'cash' for book-keeping purposes. The majority of payments and receipts are made through a bank, i.e. by cheque, credit transfer, direct debit or banker's order. However, some payments and receipts may be made in cash, or even 'kind' when goods are exchanged (this is known as contra-accounting and is covered at a later stage), which means that separate columns of figures must be maintained for each of the different methods of payment. The 'Bank' column of figures must **only** contain amounts of payment from the bank or into the bank.

For simplicity, examples at this stage will include only a 'Bank' column, without further analysis.

Cash book for Mr F.A.T. Bacon, December 19XX

Payments

	Date	Detail (from cheque counterfoils & E.M.E.B. budget account).	Bank amount	Cheque number
1	2.12.	Lance Tye., vet.	148√	019
2	4.12.	Burgess & Co., machinery reps.	263√	020
3	6.12.	Millers Ltd., pig feed	970√	021
4	6.12.	Cashed cheque wages & private	441√	022
5	27.12.	Crossroads Garage, petrol	96	023
6	27.12.	B.T., telephone	120	024
7	28.12.	E.M.E.B., electricity (Budget a/c)	136√	D.D.
8		Total December Payments	2,174	

Receipts

	Date	Detail (from paying-in book & Customs & Excise notification).	Bank amount
1	4.12.	Meatco., pigs	1,200√
2	11.12.	Drapers, wheat	1,450√
3	19.12.	Fyne Pork Ltd, cull sows	180√
4	27.12.	M & B, pigs	1,025
5	29.12.	VAT November claim	72√
6		Total December Receipts	3,927

CASH BOOK AND BANK STATEMENT – RECONCILIATION

Any form of book-keeping must always be in agreement with the bank statements and the account which shows this to be so is known as the bank reconciliation. It has been stated earlier in this chapter that the sequence of entries on a bank statement will differ from those in a cash accounting system. The accounting system will normally be ahead of, or more up-to-date than, the bank statement because payments and receipts by and to the firm are entered directly into the books. The bank does not enter them until the recipient has paid them into their account and they have passed via the Central Clearing House back to the drawer's account, and this process alone takes several days.

A bank reconciliation achieves two main purposes. Firstly, it checks the accuracy of entries and, secondly, it updates the bank statement to show the consequence of all payments from and into the bank, even though they are not yet shown on the bank statement.

Bank reconciliation: Mr F.A.T. Bacon. December 19XX

Procedures:

1. Check the list of cheques written and recorded in the cash book in the previous month which did not appear on the previous month's bank statement; tick those which appear on the December bank statement. Tick both list and bank statement.

2. Check each entry in the December cash book against the bank statement. Tick the cash book and the statement when entries appear on both and are identical.
3. Make a new list of unpresented cheques – those which appear in the cash book but not on the bank statement. Include any cheques from the previous month which are still unpresented.

4a. From cash book totals calculate the difference between payments and receipts for the month, i.e. the **cash flow**.

b. Take the 'updated' true bank balance from the previous month (November in this case) and add or subtract the cash flow for the month. The resulting figure is the **balance as per cash book.**

5. Calculate the true bank balance for the end of December, taking the balance shown on the statement and adding or subtracting

Calculations and detail

Unpresented cheques November 19XX

	Payments amount	Cheque no.	Receipts amount
	197	014	
	237	017	
	460	018	
Total	894		—

Anybank

Bank Statement				31st December 19XX
Details	Payments	Receipts	Date	Balance
Balance Forward			30 Nov	4641
Counter Credit		1200√	4 Dec	5841
000014	197√		6 Dec	
000019	148√		6 Dec	
000017	237√		6 Dec	
000018	460√		6 Dec	
000022	441√		6 Dec	4358
Counter Credit		1450√	11 Dec	
000021	970√		11 Dec	4838
000020	263√		12 Dec	4575
Counter Credit		180√	19 Dec	4755
E.M.E.B. D.D.	136√			
Customs and and Excise		72√	29 Dec	4691

Unpresented cheques December 19XX

	Payments amount	Cheque no.	Receipts amount
	96	023	1,025
	120	024	—
Total	216		1,025

Cash book December 19XX

	£
Payments	2,174 –
Receipts	3,927 +
Cash flow December	1,753 +
Bal. b.f. from November	3,747 +
Balance as per cash book	5,500 +

Bank statement December 19XX

	£
Balance at 31 Dec.	4,691 +
Unpresented payments	216 –
Unpresented receipts	1,025 +
Balance as per bank statement	5,500 +

unpresented cheques as appropriate. If the bank is overdrawn, unpresented payment cheques will be added and unpresented receipt cheques will be deducted. If there is 'money in the bank' then unpresented payment cheques are deducted and receipt cheques are added. Use plus or minus signs and compute as appropriate.

The resulting figure is known as **balance as per bank statement** and it must agree, or reconcile, with the balance as per cash book.

CASH ANALYSIS: METHOD

A record which is continuous throughout the year is clumsy and errors may go uncorrected. It is preferable, therefore, that the business year is divided into suitable periods, monthly being the most popular. Payment and receipt for each individual month are entered, totalled, balanced and transferred to a summary sheet. (Figure 4.1)

Figure 4.1 Cash analysis

Payments for December 19××

D	Date		Bank	Cheque No	Feed	Vet	Pigs	Seed	Fert	Spray	Wages	Power Mach	Rent	Admin Sundry	VAT	Private
1	2.12	Lance Tye Vet	148	019		129									19	
2	4.12	Burgess & Co Mach Reps	263	020								229			34	
3	6.12	Millers Ltd Pig Feed	970	021	970											
4	6.12	Cashed Cheque Wages Priv	441	022							341					100
5	27.12	Crossroads Garage Petrol	96	023								84				12
6	27.12	BT Telephone	120	024										105		15
7	28.12	EMEB Electricity	136	DD								136				
8			2,174		970	129					341	449		105	53	127

← 2,174 →

Receipts for December 19××

D	Date		Bank	Cheque No	Fat Pigs	Cull Sows	Wheat	Barley	Sundry	VAT	Private
1	4.12	Meat Co Pigs	1,200		1,200						
2	11.12	Drapers Wheat	1,450				1,450				
3	19.12	Fyne Pork Ltd Culls	180			180					
4	27.12	M&B Pigs	1,025		1,025						
5	29.12	VAT November Claim	72							72	
6			3,927		2,225	180	1,450			72	

← 3,927 →

Summary sheet

A summary sheet is identical in layout and column headings to the cash analysis. Cash analysis column totals for each month are transferred to the summary sheet at month end. This results in twelve sets of monthly column totals which are then added vertically to give column totals for the year. (Figure 4.2)

Coding

It is a legal requirement for tax purposes that invoices are stored in such a way that individual documents may easily be located. Efficient filing also assists administration and helps to avoid frustration and wasted time. A suitable system must therefore be devised and implemented.

A popular method involves numbering or giving a letter identification to monthly pages. This number or letter is then used together with the line number on which the payment or receipt is recorded to identify documents. The reference, e.g. C10, is written in coloured ink in the corner of each cheque counterfoil and each document so that it appears at the top right hand corner in the filing system. This ensures rapid location of individual documents.

Bank column: procedure

Record detail from cheque counterfoils and paying-in book counterfoils, entering page and line codes on the counterfoils. All counterfoil detail available for the month should preferably be entered before commencing analysis. By this means, all information is transferred from fiddly counterfoils on to a single page, which makes life much easier when sorting invoices into payment and receipt order.

Completion of the bank column ahead of analysis is also useful when checking on the bank situation with a bank reconciliation calculation, at month end or any time within the month.

Analysis

After entry in the bank column, the amount is entered again, either in one analysis column or divided between several. The amount(s) in the analysis columns must always be equal to the bank column. E.g. if a payment of £500 is for something which is zero rated for VAT purposes, perhaps food for dairy cattle, then £500 will be entered under 'Bank' and again under 'Dairy concentrate'. If a payment of £230 including £30 VAT is for machinery repairs, then £230 will be entered under 'Bank', £200 under 'Machinery costs' and £30 under 'VAT to supplier'. The effect of entering contra accounts is covered at a later stage.

VAT

Value Added Tax must be separated from payments and receipts in the cash analysis. At least one VAT column must be included in both the payment and the receipt pages although two columns in each is the preference of some, one to show VAT paid on purchases and one to show VAT paid to Customs and Excise. On the receipts page, one column would show VAT on sales and the other VAT received from Customs and Excise.

Position of the VAT column(s) on the cash analysis page is a matter of personal preference. They are commonly placed either next to the payment column or at the far side of the page.

VAT amount

The amount of VAT analysed and entered in the VAT column is **always** the amount shown on the invoice, regardless of discounts or credit charges. Analyse the VAT amount first and the remainder of the payment or receipt will then be entered in one or more of the other analysis columns.

Note that the VAT amount on a less detailed invoice may not be shown, but will have to be calculated.

Multiple invoices

When one payment covers a number of items shown on separate invoices and there is a mixture of categories of items bought or sold, it is helpful to produce a brief preliminary analysis of that payment. This may be done by determining the range of categories, and hand drawing suitable columns on the front invoice; individual items are then entered, with VAT analysed to a VAT column. Totals for each category are then entered in the appropriate cash analysis columns opposite the bank payment. Check before entry that the sum of the preliminary analysis does agree with the total payment; it is easy to make mistakes here.

MONTH END PROCEDURE

When entries in the cash analysis are completed for the month, or other accounting period, every column must be totalled. Since payments or receipts are first entered in the 'Bank' (or 'Cash' or 'Contra') column and then again in the analysis columns, it follows that the 'Bank' column total (plus cash and contra columns when used) must balance with, or be equal to, the sum total of all the analysis columns.

Bank (plus Cash plus Contra) = sum of the analysis columns

If the two totals do not balance, there is an error in the book-keeping. Find the error by the procedure on page 34, preferably using a printer/calculator.

Figure 4.2 Summary sheets

Year Ending December 19×× Payments

Date	Detail	Bank	Cheque No	Feed	Vet	Pigs	Seed	Fert	Spray	Wages	Power Mach	Rent	Admin Sundry	VAT	Private
	Jan cash anal b.f.	5,429		1,905	123			2,100		480	68		5	248	500
	Feb cash anal b.f	3,339		1,727	130		270			480	200		14	18	500
	Mar cash anal b.f.	4,076		1,450	112	220				560	132	1,050	37	15	500
	Apr cash anal b.f.	3,145		1,763	97					573	160		29	23	500
	May cash anal b.f.	3,561		1,916	86				370	580	25		36	48	500
	June cash anal b.f.	3,824		1,733	93			760		568	75		3	92	500
	July cash anal b.f.	3,140		1,680	106					536	280		22	16	500
	Aug cash anal b.f.	3,198		1,837	119					570	146			26	500
	Sept cash anal b.f.	4,851		1,985	112	450				563	163	1,050	9	19	500
	Oct cash anal b.f.	3,114		1,736	103					516	97		140	22	500
	Nov cash anal b.f.	5,371		1,868	139		650		350	374	114		25	51	1,800
	Dec cash anal b.f.	2,174		970	129					341	449		105	53	127
	Total for year	45,222		20,570	1,349	670	920	2,860	720	6,141	1,909	2,100	425	631	6,927

⟵ 45,222 ⟶

Year Ending December 19×× Receipts

Date	Detail	Bank	Cheque No	Fat Pigs	Cull Sows	Wheat	Barley	Sundry	VAT	Private
	Jan cash anal b.f.	5,688		4,460		1,120			108	
	Feb cash anal b.f.	4,415		4,120				87	208	
	Mar cash anal b.f.	6,903		4,355		2,400			18	130
	Apr cash anal b.f.	3,910		3,791	104				15	
	May cash anal b.f.	3,152		3,104				25	23	
	Jun cash anal b.f.	2,850		2,760	45				45	
	Jul cash anal b.f.	2,013		1,865				96	52	
	Aug cash anal b.f.	3,799		1,455	68		2,250		26	
	Sep cash anal b.f.	6,750		3,100	98		3,393		19	140
	Oct cash anal b.f.	4,275		3,420		810		23	22	
	Nov cash anal b.f.	6,740		3,220		3,520				
	Dec cash anal b.f.	3,927		2,225	180	1,450			72	
	Total for year	54,422		37,875	495	9,300	5,643	231	608	270

←——— 54,422 ———→

Cash analysis

a. Check arithmetic in all vertical columns.
b. Check every line horizontally to ensure that payment or receipt is equal to analysis on that line. Lay a ruler below the line to ensure that you refer only to one line at a time.
c. Check arithmetic for summation of column totals.

If errors have not been discovered at this stage then

a. Take a break and follow the above procedure again later.
b. Get someone else to do the check – a fresh eye often spots an error which your own eye passes over.
c. Get someone to read the figures aloud while you key them into the calculator – it is quite common for your eye to get used to seeing a figure but reading it differently!

Good hunting!

Transfer to summary sheet

When each completed month is balanced, i.e. bank (cash, contra) = sum of analysis columns, the column totals may be transferred to the summary sheet described above. Double check the transferred figures to ensure there are no transposition errors.

YEAR END PROCEDURE

When all of the separate months of cash analysis totals have been transferred, the summary sheet columns must then be totalled and balanced with the same procedure as for monthly cash analysis. The column totals represent total payment and total receipt for the year in each category. (Figure 4.2)

Year end adjustments

The information gathered so far represents only payments and receipts for the year, but this is not necessarily a complete picture of expenditure and revenue. It is normal for businesses to have a list of unpaid bills at the end of the financial year, owed to creditors and owed by debtors, which have not been entered in the cash analysis because they have not been paid. Unpaid bills are just as much a part of expenditure and revenue as those which have been paid and, therefore, must be included in the accounts for the year.

Unpaid bills owing at the end of a financial year will be entered in the cash analysis for the following year when paid. They are entered in this way because the 'Bank' column must include all bank payments and receipts as they occur, but the problem is that in accounting terms they belong to the previous year.

We may see from this that a cash analysis record includes creditors and debtors from the previous year's trading. They do not belong to the current year's trading and so they must be deducted from totals at the year end. Creditors and debtors outstanding at the end of the financial year must be added to cash analysis totals since they represent part of the trading for the year.

To summarise

Opening creditors and debtors are deducted from the cash analysis totals at the year end.
Closing creditors and debtors are added to the cash analysis totals at the year end.
Method:

1. List creditors and debtors for each year end, recording individual invoices with detail of date, supplier/customer, description of goods, invoice amount, VAT, amount excluding VAT. (Figure 4.3)

Figure 4.3 Year end adjustments

At 31st December 19XX

	Invoice date	Invoice no	Name of firm	Description of goods	Invoice amount	VAT	Invoice exc VAT
Creditors	2.12.XX	6442	Drapers	Pigfood	1,968	–	1,968
	4.11.XX	A141	Growell	Fertiliser	760	99	661
	8.12.XX	2022	Tractorco	Spares	820	106	714
		Total creditors at 31.12.19XX			3,548	205	3,343
Debtors	10.12.XX	–	M.&B.	Pigs	970	–	970
	12.12.XX	–	Graincorp	Wheat	1,345	–	1,345
	14.12.XX	–	J. Jones	Barley	850	–	850
	Dec Cl		C.&E.	VAT claim	201	201	–
		Total debtors at 31.12.19XX			3,366	201	3,165

At 1st January 19XX

	Invoice date	Invoice no	Name of firm	Description of goods	Invoice amount	VAT	Invoice exc VAT
Creditors	6.12.XX	10114	Drapers	Pigfood	1,672	–	1,672
	9.12.XX	7652	Growell	Fertiliser	470	61	409
	14.12.XX	4041	Tractorco	Spares	570	74	496
		Total creditors at 1.1.19XX			2,712	135	2,577
Debtors	6.12.XX	–	A. Farmer	Store pigs	320	–	320
		–	Cornco	Wheat	1,120	–	1,120
	Dec Cl	–	C.&E.	VAT claim	108	108	–
		Total debtors at 1.1.19XX			1,548	108	1,440

Figure 4.4 Summary analysis sheets including year end adjustments

Year Ending December 19XX Payments

Date	Detail	Bank	Cheque No	Feed	Vet	Pigs	Seed	Fert	Spray	Wages	Power Mach	Rent	Admin Sundry	VAT	Private
	Jan cash anal b.f.	5,429		1,905	123			2,100		480	68		5	248	500
	Feb cash anal b.f	3,339		1,727	130		270			480	200		14	18	500
	Mar cash anal b.f.	4,076		1,450	112	220				560	132	1,050	37	15	500
	Apr cash anal b.f.	3,145		1,763	97					573	160		29	23	500
	May cash anal b.f.	3,561		1,916	86				370	580	25		36	48	500
	June cash anal b.f.	3,824		1,733	93			760		568	75		3	92	500
	July cash anal b.f.	3,140		1,680	106					536	280		22	16	500
	Aug cash anal b.f.	3,198		1,837	119					570	146			26	500
	Sep cash anal b.f.	4,851		1,985	112	450				563	163	1,050	9	19	500
	Oct cash anal b.f.	3,114		1,736	103					516	97		140	22	500
	Nov cash anal b.f.	5,371		1,868	139		650		350	374	114		25	51	1,800
	Dec cash anal b.f.	2,174		970	129					341	449		105	53	127
	Total for year	45,222		20,570	1,349	670	920	2,860	720	6,141	1,909	2,100	425	631	6,927
	Less O. creditors			1,672				409			496			135	
				18,898				2,451			1,413			496	
	Plus C. creditors			1,968				661			714			205	
	Expenditure			20,866	1,349	670	920	3,112	720	6,141	2,127	2,100	425	701	6,927

Year Ending December 19XX Receipts

Date	Detail	Bank	Cheque No	Fat Pigs	Cull Sows	Wheat	Barley	Sundry	VAT	Private
	Jan cash anal b.f.	5,688		4,460		1,120			108	
	Feb cash anal b.f.	4,415		4,120				87	208	
	Mar cash anal b.f.	6,903		4,355		2,400			18	130
	Apr cash anal b.f.	3,910		3,791	104				15	
	May cash anal b.f.	3,152		3,104				25	23	
	Jun cash anal b.f.	2,850		2,760	45				45	
	Jul cash anal b.f.	2,013		1,865				96	52	
	Aug cash anal b.f.	3,799		1,455	68		2,250		26	
	Sep cash anal b.f.	6,750		3,100	98		3,393		19	140
	Oct cash anal b.f.	4,275		3,420		810		23	22	
	Nov cash anal b.f.	6,740		3,220		3,520				
	Dec cash anal b.f.	3,927		2,225	180	1,450			72	
	Total for year	54,422		37,875	495	9,300	5,643	231	608	270
	Less O. debtors			320		1,120			108	
				37,555		8,180	5,643		500	
	Plus C. debtors			970		1,345	850		201	
	Revenue			38,525	495	9,525	6,493	231	701	270

2. Analyse creditors and debtors to appropriate columns on the cash analysis summary sheet, but do not enter total amounts in the bank column. One set of creditors or debtors should be entered on one line only across the page. (Figure 4.4, pages 36–37)

Transfer to year end accounts

Adjusted amounts, which are called **expenditure** on the payments page and **revenue** on the receipts page, are transferred as appropriate to the profit and loss account (most items), depreciation account (capital purchases of equipment and machinery) or capital account (private payments and receipts).

VAT at year end

The adjusted totals of VAT paid on purchases and to Customs and Excise must balance with, or be equal to, VAT received from sales and from Customs and Excise. The adjusted VAT totals are not transferred to other accounts.

In Chapter 5 the books for Mr F.A.T. Bacon are taken forward to produce a set of year end accounts.

CHAPTER 5

From cash analysis to year end accounts

The cash analysis for Mr F.A.T. Bacon which was completed in the previous chapter is now taken forward in the preparation of year end accounts.

The balance sheet showing the state of the business at the start of the year is considered first of all. This was produced at the end of the previous financial year but it also becomes the opening balance sheet for the year under consideration. Following this, we look to the cash analysis again and extract figures for transfer to profit and loss account, depreciation account and capital account.

Balance sheet detail is taken from the schedule of valuations, the debtor and creditor list, and from the adjusted bank statement.

Balance Sheet for F.A.T. Bacon as at 1 January 19XX

Fixed assets	£	£
Sows	3,500	
Boars	480	3,980
Tractors and machinery	14,700	
Fixed equipment	3,650	18,350
Current assets		
Pigs	8,000	
Wheat	5,600	
Barley	4,200	
Feeding stuffs	1,200	
Fertiliser	850	
Seed	630	
Sundry	420	
Straw	720	
Growing crops	1,600	23,220
Sundry debtors		1,548
		47,098

Current liabilities	£	£
Sundry creditors	2,712	
Bank overdraft	3,700	
Loan	1,000	7,412
Long term liabilities		
Capital		39,686
		47,098

THE PROFIT AND LOSS ACCOUNT

The stage has now been reached when information is gathered together to produce the profit and loss account for the year. The following detail is required.

Valuations at start of year 1.1.XX
Valuations at end of year 31.12.XX
Expenditure for year (from cash analysis, Figure 5.1, page 42)
Revenue for year (from cash analysis, Figure 5.1, page 43)
Depreciation of machinery and equipment

Profit and Loss Account for year ending 31 December 19XX

Valuation at 1 Jan. 19XX	£	£	*Valuation at 31 Dec. 19XX*	£	£
Sows	3,500		Sows	3,950	
Boars	480		Boars	650	
Pigs	8,000		Pigs	9,100	
Wheat	5,600		Wheat	6,200	
Barley	4,200		Barley	4,600	
Feeding stuffs	1,200		Feeding stuffs	1,400	
Fertiliser	850		Fertiliser	900	
Seed	630		Seed	700	
Sundry	420		Sundry	470	
Straw	720		Straw	680	
Growing crops	1,600	27,200	Growing crops	1,650	30,300
Expenditure			*Revenue*		
Feeding stuffs	20,866		Fat pigs	38,525	
Vet. and med.	1,349		Cull sows	495	
Pigs	670		Wheat	9,525	
Seed	920		Barley	6,493	
Fertiliser	3,112		Sundry	231	55,269
Spray	720				
Wages	6,141				
Power and mach.	2,127				
Rent	2,100				
Admin. and sundry	425	38,430			
Depreciation					
Tractors and mach. £14,700 @ 20% p.a.	2,940				
Fixed equipment £3,650 @10% p.a.	365	3,305			
		68,935			
Profit for year		16,634			
		85,569			85,569

THE YEAR END BALANCE SHEET

Having produced the profit and loss account, the balance sheet must now be drawn up, followed by the capital account.

Balance Sheet for F.A.T. Bacon as at 31 December 19XX

Fixed assets	£	£	*Current liabilities*	£	£
Sows	3,950		Sundry creditors	3,548	
Boars	650	4,600	Loan	1,000	4,548
Tractors and machinery	11,760				
Fixed equipment	3,285	15,045	*Long term liabilities*		
			Capital		49,663
Current assets					
Pigs	9,100				
Wheat	6,200				
Barley	4,600				
Feeding stuffs	1,400				
Fertiliser	900				
Seed	700				
Sundry	470				
Straw	680				
Crowing crops	1,650	25,700			
Sundry debtors		3,366			
Bank		5,500			
		54,211			54,211

THE CAPITAL ACCOUNT

The capital account shows additions to and deductions from opening capital. The amount calculated for capital at the year end must agree with the balance. It is common to include this calculation in the balance sheet but it is shown separately here in order to ensure that the two stages are clearly understood.

Capital Account 31 December 19XX

	£
Capital at 1 Jan. 19XX	39,686
Add profit for year	16,634
Private receipts	270
	56,590
Less drawings	6,927
Capital at 31 Dec. 19XX	49,663

Figure 5.1 Summary sheets, showing allocation to appropriate accounts

Year Ending December 19XX Payments

Date	Detail	Bank	Cheque No	Feed	Vet	Pigs	Seed	Fert	Spray	Wages	Power Mach	Rent	Admin Sundry	VAT	Private
	Jan cash anal b.f.	5,429		1,905	123			2,100		480	68		5	248	500
	Feb cash anal b.f	3,339		1,727	130		270			480	200		14	18	500
	Mar cash anal b.f.	4,076		1,450	112	220				560	132	1,050	37	15	500
	Apr cash anal b.f.	3,145		1,763	97					573	160		29	23	500
	May cash anal b.f.	3,561		1,916	86				370	580	25		36	48	500
	June cash anal b.f.	3,824		1,733	93			760		568	75		3	92	500
	July cash anal b.f.	3,140		1,680	106					536	280		22	16	500
	Aug cash anal b.f.	3,198		1,837	119					570	140			26	500
	Sep cash anal b.f.	4,851		1,985	112	450				563	163	1,050	9	19	500
	Oct cash anal b.f.	3,114		1,736	103					516	97		140	22	500
	Nov cash anal b.f.	5,371		1,868	139		650		350	374	114		25	51	1,800
	Dec cash anal b.f.	2,174		970	129					341	449		105	53	127
	Total for year	45,222		20,570	1,349	670	920	2,860	720	6,141	1,909	2,100	425	631	6,927
	Less O. creditors			1,672				409			496			135	
				18,898				2,451			1,413			496	
	Plus C. creditors			1,968				661			714			205	
	Expenditure			20,866	1,349	670	920	3,112	720	6,141	2,127	2,100	425	701	6,927
				P&L	P&L	P&L	P&L	P&L	P&L	P&L	P&L	P&L	P&L	V/Ac	Cap A/c

Year Ending December 19XX Receipts

Date	Detail	Bank	Cheque No	Fat Pigs	Cull Sows	Wheat	Barley	Sundry	VAT	Private
	Jan cash anal b.f.	5,688		4,460		1,120			108	
	Feb cash anal b.f.	4,415		4,120				87	208	
	Mar cash anal b.f.	6,903		4,355		2,400			18	130
	Apr cash anal b.f.	3,910		3,791	104				15	
	May cash anal b.f.	3,152		3,104				25	23	
	Jun cash anal b.f.	2,850		2,760	45				45	
	Jul cash anal b.f.	2,013		1,865				96	52	
	Aug cash anal b.f.	3,799		1,455	68		2,250		26	
	Sep cash anal b.f.	6,750		3,100	98		3,393		19	140
	Oct cash anal b.f.	4,275		3,420		810		23	22	
	Nov cash anal b.f.	6,740		3,220		3,520				
	Dec cash anal b.f.	3,927		2,225	180	1,450			72	
	Total for year	54,422		37,875	495	9,300	5,643	231	608	270
	Less O. debtors			320		1,120			108	
				37,555		8,180	5,643		500	
	Plus C. debtors			970		1,345	850		201	
	Revenue			38,525	495	9,525	6,493	231	701	270
				P&L	P&L	P&L	P&L	P&L	V/Ac	Cap/Ac

This completes the accounts for the year. It is important that the complete procedure is understood and remembered; it helps when correcting errors.

CHECK YOUR WORK

It is not uncommon to find that the accounts do not balance and of course the checking procedure must then be followed in order to discover errors.

Procedure

1. Check that the balance as per cash book reconciles with the balance as per bank statement.
2. Sum of bank, contra and cash columns must reconcile with sum of analysis columns monthly and for the year (before year end adjustments).
3. Total payments contra column must be identical to total receipts contra column.
4. Opening debtors and creditors in cash analysis adjustments must be identical in amount to debtors and creditors in the opening balance sheet.
5. Closing debtors and creditors in cash analysis adjustments must be identical in amount to debtors and creditors in the closing balance sheet.
6. The adjusted VAT columns must reconcile, i.e. total VAT paid out must equal total VAT received.
7. The opening and closing valuations which appear in the profit and loss account must agree with those which appear in the appropriate balance sheets.
8. Check arithmetic of cash analysis adjustments.
9. Check depreciation calculations.
10. Check that figures have been transcribed from cash analysis to profit and loss correctly.

If still not balanced, repeat procedure and check arithmetic of the year end accounts.

CHAPTER 6

Double entry book-keeping

A basic concept underlying double entry book-keeping is that for every transaction there are two parties, the giver and the receiver. Further to this, the giving/receiving of goods is separated from the giving/receiving of money with which settlement is made – they are seen as two separate transactions which balance each other out.

Double entry is so named because each transaction is entered twice, once to show how it affects the business and once to show how it affects the other party – customer or supplier. By this means a complete and up-to-date picture of all purchases, sales, debtors and creditors is maintained.

Perhaps the most important difference between single entry and double entry book-keeping is that whereas entries are made only when payment occurs in single entry, in double entry all transactions are entered on receipt of or issue of invoice. Hence there is an up-to-date record of unpaid bills owed to and by the business and a true record of revenue and expenditure.

DEBTORS AND CREDITORS

Beginners at book-keeping often find it difficult to remember who are the debtors and who are the creditors when endeavouring to make entries in the correct place! The party which has **received** money or goods is **indebted**, is therefore a **debtor** and an entry to the value of their indebtedness is entered in their account.

Likewise, the party which has **given** money or goods is **credited** and is therefore a **creditor**.

We have seen that for every transaction there is a receiver and a giver. This means that in every case there is a debtor and a creditor for whom entry must be made in the appropriate account. The amount given is always equal to the amount received, which means that the books will always balance, or debits will equal credits, if entries have been made correctly.

In the example below we find that Customer 'A' purchases a gate on

2 June on credit. At the end of the month, he pays his bill. In this case 'Own Business' gave the gate and so was credited; Customer 'A' received the gate and so was debited. When Customer 'A' paid his bill he was credited, since in this case he gave, and 'Own' bank book has a debit entry because the business received and so was indebted.

Customer 'A' account

Dr		£	Cr		£
2 June	1 Gate	100	20 June	to Bank	100

'Own' sales account

Dr		£	Cr		£
			2 June	1 Gate (Customer 'A')	100

'Own' bank/cash book

Dr		£	Cr		£
20 June	Customer 'A'	100			

We see that in this case there has been a debit and credit entry for each of the two 'give' and 'receive' transactions. The total of the debits balances with, or is equal to, the total of the credit. No matter what volume of entries is made, this principle must always apply and the books must always balance – or the reason be found why not!

DIFFERENT CLASSES OF ACCOUNT

Although the basic principles of double entry book-keeping as explained above always apply, the issue is a little more complex in that the accounts are classified in a particular way. We need to understand why the different classes exist and what they are called.

There are three classes of accounts, personal accounts, real accounts and nominal accounts. In each case the accounts are seen as receiving or giving value for which they are debited or credited.

Personal accounts

Personal accounts are a record of transactions with other businesses, whether they be suppliers or customers. Personal, because they are associated with people who represent the other businesses. When another business supplies goods or services, they are credited and become creditors to our business, when they receive goods or services from our business they become debtors.

Real accounts

Real accounts record the giving and receiving of real, tangible things like machinery, equipment and money. The accounts 'receive' the assets which are held at the start of the business year and then purchases and sales are recorded as normal. Depreciation of machinery and equipment is entered as a credit because that account which represents the machinery is 'giving' value which causes the machinery to be worn out and fall in value.

For example, Bank account at the start of the year has a debit balance of £10,000 because the account has received and holds the money on behalf of the business. A cheque for £500 is paid out to a supplier, Macspares Ltd (a personal account) in settlement of their account.

(*Real Account*)

Bank Account

Dr		£	Cr		£
6 April	Bal. b.f.	10,000	10 April	Macspares Ltd	500

(*Personal Account*)

Macspares Ltd Account

Dr		£	Cr		£
10 April	Bank	500	6 April	Bal. b.f. (An 'opening creditor')	500

At the end of the year the machinery account which recorded the purchase of a £5,000 machine shows depreciation of £1,000.

(*Real Account*)

Machinery Account

Dr		£	Cr		£
20 April (start of year)	New trailer	5,000	5 April (end of year)	Depreciation	1,000
				Debit balance being written down value c.f.	4,000
		5,000			5,000

Nominal accounts

As the name implies, these are accounts which are given names and which are created for our own purposes. Nominal accounts record the main business activities of buying and selling (other than capital equipment and machinery) and the provision and receipt of services. They do in fact record the activities which primarily influence profit or loss for the trading period.

The range of accounts created must be determined by establishing **what information is required by the business.** For a simple business with only one line of sales or service, the solution is simple. When there are a number of products and services sold by the business, with a related broad range of purchases or inputs, then accounts must be created in order to reflect this. A conglomerate account with no detail of sales or purchases leads to successful activities masking the failures. The approach to creating the accounts should be identical to that followed in determining headings for cash analysis in single entry book-keeping; but remember the key difference between the two systems is that in double entry, entries are made as soon as the invoice is produced, whereas in single entry, the entry in the books only occurs at the time of payment.

THE JOURNAL OR DAY BOOK

In the days of lowly paid ledger clerks the whole double entry accounting system was held together by the journal proper, a sort of 'check in' which all transactions had to go through prior to entering the account to which they belonged. This meant that there were four entries for every transaction, not just two! The journal still exists in common use but its function is limited to the entry of balances of personal and real accounts at the beginning and the end of the year – a list of assets and liabilities, in fact, the assets being the debits (because of value received) and the liabilities being the credits (because of value given by the business).

There must be an orderly system of making entries to the many accounts on a day by day basis and to achieve this, the bank account, the sales account and the purchases account are considered to be 'books of first entry'.

DESIGN OF BOOK-KEEPING STATIONERY – NOMINAL ACCOUNTS

Sales accounts only 'give' and are credited, purchase accounts only 'receive' and are debited. It follows therefore that it is clumsy and unnecessary to provide for both debits and credits in the stationery if only one side of the account is to be used. The whole system may be streamlined by adopting multi-column analysis stationery and so in fact the sales book is sales analysis (credit) and the purchase book is purchase analysis (debit). The first column records total sales or total purchases and functions as a book of first entry in place of the journal proper. The sale or purchase entry is then made again under the appropriate column heading.

Sales and purchases are entered in the sales book or purchase book **and** in the appropriate personal ledger when invoices are produced or received. When payment is made the entries are made in the bank account and in the appropriate personal account. If payment is made at the point of sale or purchase the detail is entered only in the bank account and the sales or purchase account, as appropriate.

For example, Vale Pony Supplies Ltd supply Merrivale Riding School with £100 worth of pony nuts. Sales account (Pony Nuts) is credited £100, the personal account for Merrivale Riding School is debited £100.

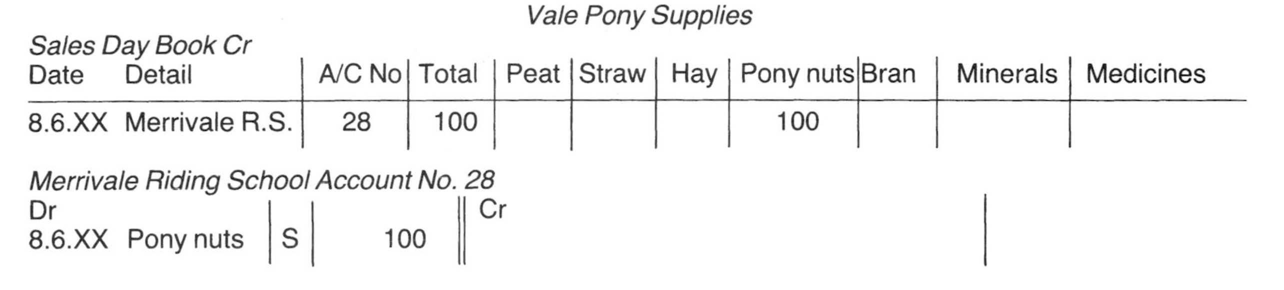

Vale Pony Supplies

Sales Day Book Cr

Date	Detail	A/C No	Total	Peat	Straw	Hay	Pony nuts	Bran	Minerals	Medicines
8.6.XX	Merrivale R.S.	28	100				100			

Merrivale Riding School Account No. 28

Dr				Cr
8.6.XX	Pony nuts	S	100	

When large numbers of customers are involved it becomes necessary to number or code the accounts to enable detail to be checked quickly and easily. In this case Merrivale Riding School Account is numbered 28 and this is entered in the reference column in the sales book. The S code in the reference column in Merrivale's account indicates that the indebtedness is from sales account. Both entries are easily cross-referenced by use of code and line numbers.

THE TRIAL BALANCE

The trial balance is a checking procedure which is followed to ensure that every debit entry has been balanced with a credit entry. This is achieved by first 'balancing off' individual accounts and then taking the resulting balances to the trial balance.

'Balancing off'

Bank Account

Dr	£	Cr	£
1 June b.f.	800	Telephone	250
4 June Sales	1,500	Diesel fuel	160
		Balance c.f.	1,890
	2,300		2,300
Balance b.f.	1,890		

The balancing off procedure often mystifies students – but there is no mystery! Just total the two sides separately and then subtract the smaller from the greater. The difference between the two figures is the **balance** which is then entered on the side of the account with the smaller amount as indicated above.

	£
Total of debit side	2,300
Total of credit side	410
Balancing amount (Dr)	1,890

The greater of the two sides always determines what the balance is – debit or credit. In the example the debit side is greater and so the residue after deducting total credit is a **debit balance**. Balances may be entered and the account ruled off as indicated or, if it is not the end of a book-keeping period (day, week or month), the balances may just be pencilled in. When entries appear on one side of the account only, either debit or credit, there is little point in balancing; just total the entries for the one side.

This is a trial balance for a simple set of books for a general and agricultural contractor.

Trial balance as at 30 June 19XX

	Dr £	Cr £
Sales and contract a/cs		
Haulage		2,250
Digger work		890
Forage harvesting		1,450
Baling		600
Purchase and expense a/cs		
Fuel and oil	670	
Repairs and spares	830	
Insurance and Road fund	550	
Wages	1,260	
Telephone and administration	440	
Sundry	270	
Sundry debtors a/c	930	
Sundry creditors a/c		650
Plant and machinery a/c	22,250	
Bank a/c	4,300	
Drawings a/c	1,250	
Capital a/c		26,910
	32,750	32,750

Each of the above debits and credits has been brought forward from the individual accounts shown on the facing page. The example is deliberately kept at a simple level so that the student may 'see the wood' rather than get lost amongst the detail or 'trees'. Check each balance in the accounts – the only way to learn is to get involved.

Personal Accounts

Tom Cobley Account

Dr	£	Cr	£
Haulage	250	Bank	250
Baling	370	Balance (Dr) c.f.	370
	620		620

Arthur Seagall Account

Dr	£	Cr	£
Forage harvesting	560		
(Dr bal. 560)			

Fuelco Account

Dr	£	Cr	£
Bank	200	Diesel fuel	450
Balance (Cr)	250		
	450		450

Partco Account

Dr	£	Cr	£
		Spares	400
		(Cr bal. 400)	

Drawings Account (for owner of the business)

Dr	£	Cr	£
Cash	500		
Cash	500		
Cash	250		
(Dr bal.)	1,250		

Capital Account

Dr	£	Cr	£
		1 June Capital b.f.	26,910

Real Accounts

Plant and Machinery Account

Dr	£	Cr	£
1 June capital	22,250		

Note Plant and machinery would be listed in a separate schedule. Investment in assets of this type is known as capital investment and is entered as such.

Bank a/c

Dr	£	Cr	£
1 June Bal. b.f.	5,100	Fuelco	200
Cobley	250	Drawings	1,250
Cash	400	Balance (Dr) c.f.	4,300
	5,750		5,750

Credit *Sales and Contract*

Date	Detail	Ref	Total	Haulage	Digger	Forage harvesting	Baling	Sundry	
	Detail of individual entries omitted. VAT is not taken into consideration at this stage.								
	TOTAL SALES		5,190	2,250	890	1,450	600	–	

Debit *Purchase and Expense*

Date	Detail	Ref	Total	Fuel oil	Repairs spares	Ins., road fund	Wages	Tel., admin.	Sundry
	TOTAL PURCHASES		4,020	670	830	550	1,260	440	270

THE PROFIT AND LOSS ACCOUNT

The abbreviated and simplified set of books considered in this chapter so far cannot be considered complete or representing the activities of a real business, but they demonstrate the basic principles of double entry. The exercise is now developed a step further by constructing a balance sheet and profit and loss account for the period.

Balance sheet and profit and loss account are unaffected by book-keeping method. Single entry or 'cash accounting' systems must be adjusted for creditors and debtors to show the true results of trading, but this is not necessary when a double entry system is followed. Since transactions are recorded in the sales and purchase books when they actually occur, a true record is provided at the period end regardless of whether accounts have been settled.

From the trial balance we extract information firstly for the profit and loss account below, followed by balance sheet and capital account.

Profit and Loss Account for A Contractor for year ending 30 June 19XX

Expenditure	£	*Revenue*	£
Fuel and oil	670	Haulage	2,250
Repairs and spares	830	Digger work	890
Insurance and road fund	550	Forage harvesting	1,450
Wages	1,260	Baling	600
Telephone and admin.	440		
Sundry	270		
	4,020		
Profit	1,170		
	5,190		5,190

Balance Sheet for A Contractor as at 30 June 19XX

Fixed assets	£	*Long term liabilities*	£
Plant and machinery	22,250	Capital	26,830
Current assets		*Short term liabilities*	
Bank	4,300	Sundry creditors	650
Sundry debtors	930		
	27,480		27,480

We now adjust the capital a/c for the effect of profit and drawings.

Capital Account as at 30 June 19XX

	£		£
Drawings	1,250	1 June Capital b.f.	26,910
Capital	26,830	30 June Profit	1,170
	28,080		28,080

If we examine the accounts we soon discover that owner's capital, or equity, has fallen but this is clearly due to the fact that drawings from the business are greater than profit for the period.

CAPITAL AND THE CAPITAL ACCOUNT

We have looked at balance sheets and considered the meaning of capital or equity. There are various interpretations of these terms and even several more terms including net capital, capital owned and net worth. Development of understanding must start from a solid baseline and so for consideration of relatively simple business situations we will regard capital as being the investment made by the proprietor(s) in the business.

The point was made in Chapter 1 that accounting processes can be seen in stages. We may now suitably review the earlier chapters and see how the stages in accounting fit together.

Stage 1

The balance sheet shows, amongst other detail, the capital invested in the business by the proprietor at a stated point in time. This information is of great importance to the owner and it serves as a measure against which success or failure may be set. Capital may be seen as a starting point for a trading period, to which profit may be added or from which loss may be deducted. Further to this, the owner may withdraw money for private use or may add to capital by transferring in private funds.

Stage 2

The book-keeping stage builds a record of revenue and expenditure plus capital purchases and sales (purchase and sale of capital items such as machinery, equipment, buildings) and of course withdrawal of money for private purposes.

Stage 3

The profit and loss account is constructed by use of information from the books for the period, from valuations at the beginning and end of the period and from the depreciation account. The profit and loss account sets the inputs against the outputs to determine profit or loss for the trading period.

Stage 4

The balance sheet at the end of the year again shows the position of the business. The capital account shows additions to and deductions from capital during the year and agrees with the closing balance sheet.

CHAPTER 7

But it is not quite so simple!

Basic principles of book-keeping have been dealt with in previous chapters but there are further issues which cause complications and so it is important that they are fully understood. Each one will be explained and demonstrated through simple examples and then integrated into a full set of books in a following chapter.

CONTRA ACCOUNTS

Barter or 'payment in kind' for goods is probably the oldest form of trading there is and it still goes on in this computerised, near twenty first-century world! Everyone is familiar with the concept of 'trading in' a vehicle when purchasing a replacement: a value is fixed for both vehicles and the replacement vehicle is paid for by handing over the old vehicle, plus any money required to make up the price of the new one. What is the point in handing over sums of money when it has to be handed straight back?

Contra simply means 'against' and is used in many contexts. Contra accounting normally takes place when there are two simultaneous transactions, with one party having a major involvement in handling the goods or services with which the transaction is concerned. A common example is when goods are sold by auction. The auctioneer does not pay over the full sale price to the seller (vendor) and then issue a bill for auction expenses – he is too canny for that! The sale document shows the full price realised and in a separate section the commission, charges and VAT are detailed, the total being deducted from the sale price to determine the amount payable to the vendor.

County Auctions Ltd. Sold on behalf of A Farmer

	£	£
4 cattle		1,200
Less		
Commission	36	
Charges	4	
VAT	6	
	46	46
Cheque		1,154

For book-keeping purposes the two sections of the document must be treated as two separate invoices. The full value of the cattle must be shown in the receipts analysis and the cost of selling must be entered as a payment – but it was paid **by contra account** (setting against) and so a 'Contra' column must be included in the cash book. The contra account must be regarded simply as a method of making payment.

Receipts analysis

Date	Detail	Bank	Contra	Milk	Calves	Cattle	Sundry	VAT
	C. Auctions (4)	1,154	46			1,200		

The full value of the cattle sold is entered in the 'Cattle' column and the amount received is split between 'Bank' and 'Contra'. The 'Contra' column simply indicates that the business has received value, in this case it is the value of the auction charges.

Payments analysis

Date	Detail	Bank	Contra	Feeds	Vet./med.	Cattle	Lst. sundry	VAT
	C. Auctions charges		46				40	6

The auctioneer's commission and charges are entered as 'Livestock sundry' costs and the VAT is entered in the 'VAT' column since it can be reclaimed. Payment has been made by contra account; part of the value of the cattle has been given in payment for the services of the auction and the VAT charged.

Contra accounting is about giving value in exchange for something of the same value and so contra payments and contra receipts are **always equal** – the same amount must always be entered in both payment and receipt records and so they balance out. Always check the total contra account at the month end – if receipt is not equal to payment there is an error and the accounts will not balance at the year end (or whenever accounts are drawn up).

CONTRA ACCOUNTS – DOUBLE ENTRY

A simple method of contra accounting in double entry is identical to that used for single entry.

Contra accounting is only necessary at the point of payment or settlement. The bank account in a double entry system records payment and receipt at the same stage and in the same way in single entry. The only difference is that the bank account is not analysed for double entry since that is done at an earlier stage in the sales and purchase accounts.

A 'Contra' column is maintained together with the 'Bank' column and the two entries are made at the same stage.

Bank account

Date	Debit		Bank	Contra	Credit		Bank	Contra
1.	Sales C. Auctions	S1	1,154	46	Purchases C. Auctions charges	P1		46
2.								
3.								

Sales account

Date	Detail		Total	Milk	Calves	Cattle	Sundry	VAT	
1.	Cattle auction	BD1	1,200			1,200			
2.									
3.									

Purchase account

Date	Detail		Total	Feed	Vet./ med.	Cattle	Lst. sundry	VAT	
1.	Cattle Auction contra charges	BC1	46				40	6	

Petty cash

The problem with petty (small) cash is that the purchases and receipts are generally insignificant on their own and they tend to be left 'until later' before a record is made.

Type and size of business will obviously determine the volume of petty cash transactions. A one man business with a low volume of petty cash transactions may manage quite satisfactorily with a simple, informal system. When a number of people are handling petty cash, or there are a considerable number of transactions, then a more formal approach is required.

The imprest system

The imprest system is based on a float, or pre-determined amount of money to be available for petty cash purposes, according to need. Cash is drawn from the float as required and when the petty cash is balanced the deficit is made up by withdrawal from the bank. The bank amount is entered in the cash analysis book and analysed in the appropriate columns from the petty cash book. When a double entry system is maintained the petty cash book may be regarded as a purchase (or sales) book in its own right or the weekly/monthly entries may be taken to the main purchase book and allocated to relevant columns or accounts.

VAT amount in less detailed invoices

Invoices given for petty cash purchases are likely to be of the **less detailed*** type on which the VAT amount is not shown. It will be necessary to calculate the VAT amount for taxable items when a less detailed invoice is given.

* See VAT General Guide published by Customs & Excise.

Calculation of VAT in less detailed invoices

The invoice total includes the price of the goods plus VAT, when it is stated on the invoice that VAT is charged. It is a legal requirement that the percentage rate of VAT charged is stated, but many firms do not bother to include this if the invoice is written by hand.

Since the invoice amount includes both the price of the goods and the VAT, it is incorrect to use the percentage rate of VAT for the calculation. The VAT fraction must be used to calculate the VAT content. The VAT fraction is determined as follows.

	£		£
Goods per	100		100
VAT chargeable at e.g. 15%	15	e.g. 17.5%	17.5
Total	115		117.5

It follows that £15 out of every £115 or £17.50 out of every £117.50 is VAT and this can be expressed as a fraction.

$$\frac{15}{115} = \frac{3}{23} \quad \text{or} \quad \frac{17.5}{117.5} = \frac{7}{47}$$

When VAT is charged at 15 per cent, 3/23 of total invoice amount is VAT and at 17.5 per cent, 7/47 of total invoice amount is VAT.

e.g. Calculation of VAT content on total amount of £46:

15% VAT $46 \times \frac{3}{23}$ = £6 VAT £40 exc. VAT

17.5% VAT $46 \times \frac{7}{47}$ = £6.85 VAT £39.15 exc. VAT

Customs and Excise publish the VAT fraction but it is helpful to know how it is determined and used.

The VAT content of petty cash items in the following example is calculated at 15 per cent using the appropriate VAT fraction.

(Imprest) Petty Cash

Debit (receipts)	£	Credit (Expenses)	Total £	VAT £	Exc. VAT £
Bal. b.f.	50.00	Car parking	2.20	0.29	1.91
		Envelopes	1.30	0.17	1.13
		Stamps	3.10	–	3.10
		Telephone	2.20	0.29	1.91
		Adhesive tape	4.20	0.55	3.65
			13.00	1.30	11.70
		Cash from bank	37.00		
		Balance c.f.	50.00		
	50.00		50.00		

Alternative – less formal with no fixed float – just 'out of pocket'.

Petty cash book

	Total £	VAT £	Exc. VAT £
Car parking	2.20	0.29	1.91
Envelopes	1.30	0.17	1.13
Stamps	3.10	–	3.10
Telephone	2.20	0.29	1.91
Adhesive tape	4.20	0.55	3.65
Cheque cashed to P.C.	13.00	1.30	11.70

The less formal approach is suitable for a situation where these minor expenses are paid 'out of pocket'. The system of drawing cash from the bank to cover the past week/month's petty cash expenses enables proper entry in the cash analysis book without the use of a cash column. It is not necessary to draw a specific amount to cover petty cash, simply to make adequate drawings for private use but analyse the petty cash as part of it.

Posting the petty cash book to cash analysis book

Cash Analysis
Payment

Date	Detail	Bank					Admin., sundry	VAT	Private drawings
	Private – P. Cash	60					11.70	1.30	47

Note: The analysis would be identical if entered in a double entry purchase book.

Analysed petty cash

When there is a large volume of varied petty cash expenses it may be desirable to extend the petty cash book with analysis columns. This is done in exactly the same way as previously considered for the cash analysis book and the column totals are carried to the cash analysis book (single entry) or the purchase book.

Petty cash receipts

If possible **all** petty cash receipts should be paid into the bank account. This ensures that there is a proper record of these receipts and the accountant will more easily convince the tax inspector of accuracy in this context.

If petty cash receipts are held to replenish the float in an imprest (or similar) petty cash system, then detail of receipts must be recorded. Use of the 'Contra' column will become necessary when receipts partially or fully offset payments – or vice versa – in order to reflect

Petty Cash

Debit	£	Credit	£
Bal. b.f.	50.00	Stationery	3.00
Timber	10.00	Taxi	5.00
Bank (cheque cashed)	15.00	Business magazine	2.00
		Petrol	15.00
		Total payments	25.00
		Balance c.f.	50.00
	75.00		75.00

Cash Analysis

Payments

Date	Detail	Bank	Contra		Petrol Car exp.		Admin., sundry	VAT	Private drawings
	Petty cash b.f.	15	10		13.04		8.70	3.26	

Receipts

Date	Detail	Bank	Contra			Timber		VAT	
	Petty cash b.f.	–	10			8.70		1.30	

full values in both payment and receipt sides of the cash analysis or sales and expense in a double entry system.

PETTY CASH – DAY TO DAY MANAGEMENT

When a petty cash float is maintained for use by a number of staff **it is essential that receipts and withdrawals of cash are recorded as they occur**. A popular method of achieving this is the use of petty cash 'slips' or 'chits', on which detail is written and the slip put in a box or file. Suitable pads of printed petty cash slips can be purchased from any business stationery supplier.

VALUE ADDED TAX

VAT regulations are constantly changed and updated and so a textbook on accounting and book-keeping is not the place to consider any more than basic principles, and even they can change from time to time. Heavy penalties can be imposed for failure to comply with regulations and so it is important that proprietors, book-keepers and accountants maintain up-to-date knowledge of the current situation by reference to information published by Customs and Excise.

Businesses which are VAT registered are entitled to reclaim VAT paid on purchases and they must charge VAT on standard rated goods and services and pay the VAT collected to Customs and Excise. However, there are exceptions to this general guide. Exempt and partially exempt businesses may not be entitled to reclaim VAT on certain goods and services. Reference to Customs and Excise publications is essential.

VAT in book-keeping and accounting

When VAT is reclaimable, as in the majority of cases, it is not a cost to the business. Likewise VAT charged on goods and services is payable to Customs and Excise and so there is no gain to the business. The main effects on the business are increased administration costs and the possibility of adverse cash flow when debtors (owing amounts which include VAT) are slow to pay.

Assuming that VAT is reclaimable, all VAT is deducted from invoices and analysed to the VAT account or column, leaving the true cost excluding VAT to be entered in the relevant account or analysis column.

The VAT charged on goods and services must be shown on fully detailed invoices and it is that amount which is analysed to the VAT column. Less detailed invoices may state the total invoice amount including VAT at the applicable rate and in this case the VAT amount must be calculated.

VAT returns and necessary recording systems

To comply with regulations book-keeping systems may be adapted or special VAT recording systems may be used. Detailed instruction is not given here because the regulations change constantly. The reader is referred to current Customs and Excise VAT publications.

DISCOUNT AND CREDIT CHARGES

Single entry book-keeping

When a single entry, cash analysis book-keeping system is used, the price paid is the final cost and so there is no point in creating a record of discounts lost or credit charges paid for normal accounting purposes. If there is evidence that considerable discounts are being lost and credit charges paid, then it may be desirable to introduce a column in the cash book for monitoring purposes. The most important action to take will be to maintain a regular check on invoice dates and the final dates for payment, after which discount is lost or credit charges become payable, and pay the bill! Lost discount can be more expensive than increasing an overdraft.

For example, say 5 per cent discount is allowed if an invoice is settled within 28 days. If the bill is settled at the end of the second month instead of the first, an additional cost of 5 per cent must be paid due to lost discount. 5 per cent for one month is equal to $12 \times 5\% = 60\%$ for one year – rather more costly than a bank overdraft!

Double entry book-keeping

When a double entry system is maintained and invoices are raised from which discount may be deducted, then it is necessary to record these adjustments. As always, this should be done simply but effectively.

For example, goods are sold by the Rustic Gates business on which 5 per cent discount is allowable if the account is settled within 28 days.

Sales Book (for Rustic Gates Ltd)

Date	Detail		Total	Field gates	Garden gates				VAT
1.6	Vale Garden Centre	L14	60.00		52.50				7.50

Vale Garden Centre (Personal Account) L14

Date	Debit	Dis. £	£	Date	Credit		Dis. £	£
1.6	Garden Gate		60.00	28.6	Cheque		2.50	57.50

Bank Account

Date	Debit	Dis. £	Bank £	Date	Credit	Dis. £	Bank £
	Bal. b.f.		1,000				
28.6	Rustic Gates L14	2.50	57.50				

Note: In a more historically 'pure' double entry system, the discount account would be separate or even developed further to 'discount given' and 'discount received' accounts, but to simplify it as in the example has the additional benefit of giving 'at a glance' customer information regarding promptness of payment and discounts allowed.

CHAPTER 8

A step further

DEPRECIATION OF CAPITAL ASSETS

The goose that laid the golden eggs unfortunately grew old and died. Likewise vehicles, machinery, equipment and buildings which contribute to the productive process all grow old and deteriorate, leading to inefficiency and costly maintenance. More formally, depreciation may be seen as a permanent fall in value and utility of an asset due to 'wear and tear' and obsolescence.

Why not just enter the purchase price of capital items in the year of purchase? This may be acceptable in the case of smaller items of equipment but it would be unreasonable to burden any one business year with the cost of an asset which still exists at the end of the year and which will continue to contribute to production for a number of years. An item of equipment such as an agricultural tractor may cost many thousands of pounds and at the end of the year will still be complete and capable of years of hard work – but the wearing and ageing process has begun. The problem faced by the accountant is how to apportion fairly the fall in value that takes place.

Depreciation – not capital allowance

It is important to recognise the difference between depreciation and capital allowance against tax. Depreciation is an 'informed guess-timate' of the fall in value of the asset over a trading period, whereas a capital allowance is that part of the cost which the Board of Inland Revenue will allow against profit. The allowance may or may not be similar in amount to the calculated depreciation.

Depreciation – methods of calculation

Rate and pattern of fall in value of an asset is dependent on the type of asset under consideration and how it is employed. A high cost, high technology asset like a tractor, a truck or a combine harvester, which may be resold at any stage of its working life, will normally fall in value by the greatest amount in its first year, followed by a gradually decreasing annual amount. This is realistic and will reflect normal personal experience of car purchase – removal from the showroom and registration alone causes the first significant fall. Buildings and fixed equipment are not normally subject to the heavy wear and tear suffered by mobile assets and they are not normally purchased with a considered option of resale in the short term – they are normally a medium to long term investment and the fall in value is normally spread evenly over the life of the asset.

Reducing balance method of depreciation

This method is normally used for vehicles, machinery and equipment. Depreciation for the first year is calculated on cost.

Machine cost	£10,000
Annual depreciation 20%	
Year I £10,000 × 20% =	£2,000 depreciation
£10,000 – £2,000 =	£8,000 written down value (or balance)

Depreciation for the second and subsequent years is calculated on the W.D.V. (written down value) calculated at the end of the previous year.

Year 2	
W.D.V. at end of year I	£8,000
Depreciate at 20%	£1,600 = Depreciation for year
W.D.V. at end of Year 2	£6,400

Year 3	
W.D.V. at end of Year 2	£6,400
Depreciate at 20%	£1,280 = Depreciation for year
W.D.V. at end of Year 3	£5,120

It will have become clear from the above example that, as the W.D.V. (or balance) reduces year by year, so consequently does the annual depreciation. Presentation of annual depreciation on a graph demonstrates clearly the pattern over the useful life of the asset. In the final years the fall in value is low and as a consequence the life of the asset could continue on paper for many years but the normal procedure is to write off the asset at the end of its estimated useful life. If it is sold for more than the current W.D.V. the margin will contribute to profit for the year.

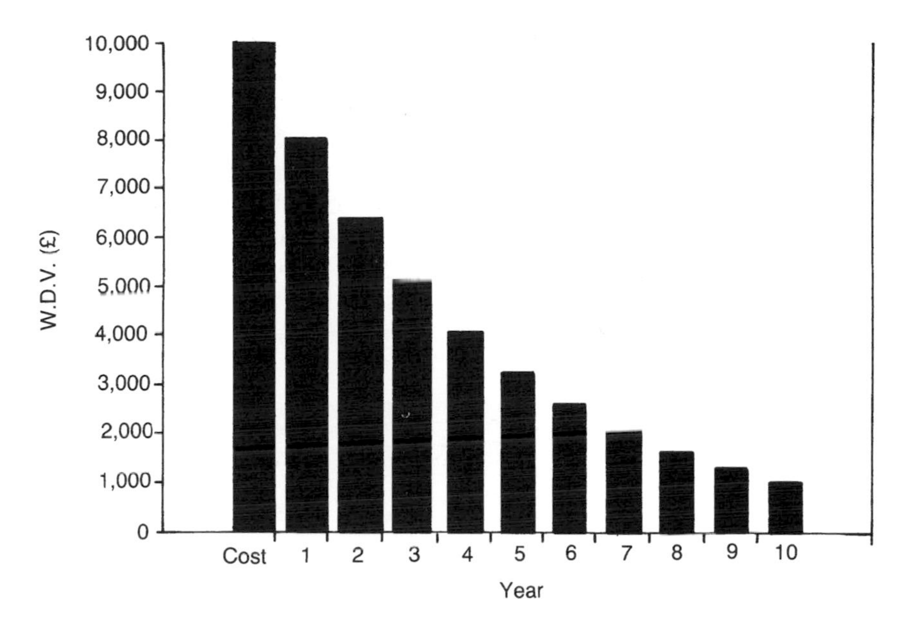

Figure 8.1 Reducing balance depreciation

Straight line method of depreciation

This method is normally used for fixed equipment and buildings. Depreciation in the first year and all subsequent years is calculated on cost and so the annual fall in value remains the constant.

Cost of asset £10,000
Estimated life 5 years
Annual depreciation ⅕/year = 20%

Depreciation Year I £10,000 × 20% = £2,000

Cost	£10,000
Less depreciation	£2,000
W.D.V. at end of Year I	£8,000

Depreciation Year 2 (cost) £10,000 × 20% = £2,000

W.D.V. b.f.	£8,000
Less depreciation	£2,000
W.D.V. at end of Year 2	£6,000

Depreciation Year 3 (cost) £10,000 × 20% = £2,000

W.D.V. b.f.	£6,000
Less depreciation	£2,000
W.D.V. at end of Year 3	£4,000

Graphic presentation of this method produces a straight line since the annual fall in value is identical throughout the life of the machine. An estimated scrap value is sometimes included in the calculation but this is difficult to foresee – and the whole cost will eventually have been included in the annual accounts so it is reasonable to write it off in the final year of depreciation, even if it is subsequently sold and the sale value contributes to profit.

Note: The rate of 20 per cent is taken simply for illustration and comparison with the previous method shown. Buildings are more likely to be depreciated at 10 per cent p.a.

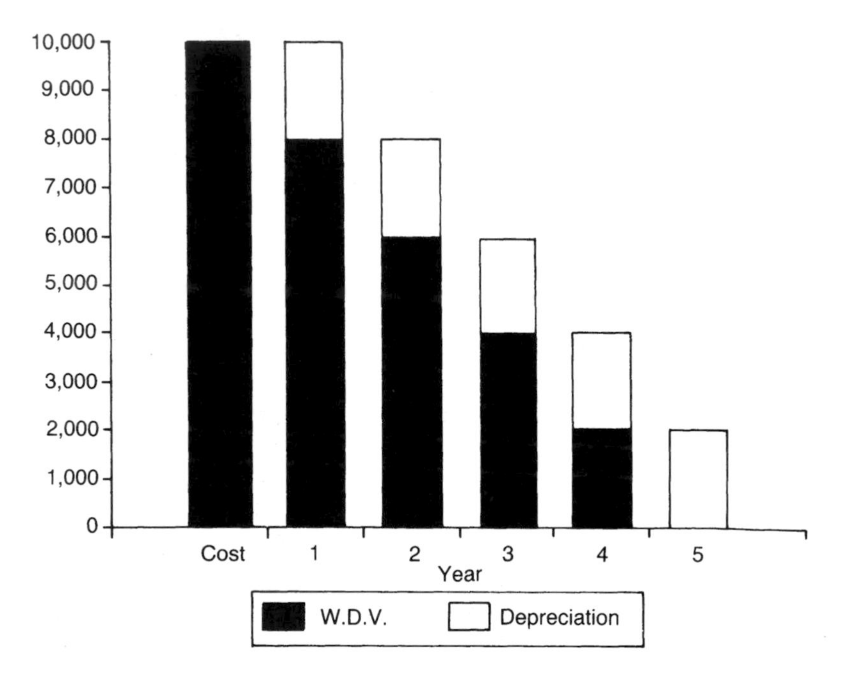

Figure 8.2 Straight line depreciation

Pool depreciation

In practice it may be convenient to pool machinery according to type, e.g. tractors, harvesting equipment, cultivating equipment etc. When this is done, individual items may be purchased and added to the pool in the year of purchase or sold and deducted from the pool.

Orderly layout of all accounting procedures is essential. The following may be found suitable for most situations when pooling takes place.

	Vehicles	*Tractors*
W.D.V. at (date – start of year)	£32,000	£44,000
Plus purchases	4,000	12,000
	36,000	56,000
Less sales	2,000	4,000
	34,000	52,000
Less depreciation @20%	6,800	10,400
W.D.V. at (date – end of year)	27,200	41,600

Register of capital assets

Calculation of depreciation on a pool basis may be considered convenient and less tiresome but it is important that a detailed register or inventory of all vehicles, tractors and equipment is maintained with purchase date and price and disposal date and price. This not only ensures that items are not overlooked in the depreciation calculation but it is also useful for insurance and security purposes – theft of machinery and equipment is by no means uncommon and it does help if you know what has gone.

VALUATIONS AND STOCK CONTROL

Perhaps one of the most difficult aspects of measuring profit by means of annual accounts is the fact that in many businesses there are material assets for which a value must be determined. Too high a value will have the effect of inflating profit (on paper) for the year, whereas too low a value will decrease the profit calculated.

There are three main classes of asset which must be valued at year end, in addition to capital assets which are subject to depreciation calculation. The three classes include goods purchased for resale, raw materials purchased for inclusion in a productive process and goods to which value has been added; the most difficult assets to value are those in the third group which are unfinished at the year end.

Goods purchased for resale

This class of goods is relatively simple to cope with since they are valued at cost. It is important that stock records are accurately maintained and that goods are resold in the sequence in which they were purchased.

If goods are held for long periods before resale then some may consider it reasonable to add storage and finance charges – but they should also look carefully at stock turnover and possibly reduce stocks held.

Raw materials

Raw materials are normally valued at cost and once more proper stock control is of utmost importance.

Unfinished goods – 'work in progress'

It is difficult to be both accurate and consistent in valuation of unfinished goods. First it is important that profits should not be anticipated by adding a profit element to unfinished goods. Valuation of unfinished goods may be restricted to the inclusion of physical inputs or, if suitable cost records are maintained, such items as labour and fuel may be included. **Consistency** is all important. It would be quite wrong, for instance, to include labour in this valuation at one year end but not at the next. An agricultural or horticultural example under this heading could include growing crops. Extensive crops such as cereals are normally valued at cost of seed, fertiliser and crop protection inputs whereas intensive horticultural crops may include other elements of cost.

Revaluation of goods

This is not a fourth classification but rather a development of the third group given above and in the agricultural context would apply to the valuation of livestock. Agriculture is unusual in that livestock may change in value during the year, either due to growth and an improvement in condition or due to a decline in condition, and so be revalued. Valuation of breeding livestock is preferably based on a standard value per head which should only be adjusted to take account of considerable inflation and/or a definite change in the quality of the herd. Value should reflect average market price less suitable market charges. It must be remembered that we are considering valuation for management accounts. Information regarding adjustment of valuations for tax accounts is available from Inland Revenue publications.

Valuation of trading livestock – those which are produced for sale rather than kept for breeding purposes – may reflect market values if they are 'finished' and ready for sale, otherwise they should be valued at cost. If there are insufficient records available to allow calculation of cost, then a suitable method is to value at 75 per cent of market value.

Valuation of harvested crops in store

It is realistic to value harvested crops at near market value less suitable selling and transport charges. It is important that a rather pessimistic view of price is taken rather than an optimistically high price, particularly if there are fluctuations in the market.

Herd basis valuation of breeding livestock (for tax purposes)

Farmers may opt to value breeding livestock on the herd basis, which in simple terms has the effect of removing the herd from the profit and loss account and treating it as a capital asset – herd depreciation and replacement costs are included in the account by entering cull sales and rearing costs. This book is not aimed at the production of tax accounts, so this subject is only briefly touched upon. Inland Revenue publications will provide more detailed information.

Schedule of valuations

Full detail of all classes, age groups and numbers of stock should be recorded with individual values. Similarly all crops and stores should be suitably classified with numbers or quantity and unit price. Valuations should be carried out on the same date each year – or near the same date at least. Perhaps most important of all is the need for **consistency** of method. It is only by compiling consistently accurate and detailed records of valuations that, together with other records, it becomes possible to produce accurate enterprise costings and gross margins.

Stock check

Numbers or quantities of live and dead stock in valuations must be checked for accuracy against stock records. A stock record will include opening stock, plus additions to stock, minus deductions from stock (for any purpose). The balance will show current or end of year stock held, which must be checked against physical stock to ensure accuracy and to check on losses through waste or theft.

Professional valuers

Many farmers engage a professional valuer to carry out year end valuations. The information provided may be a conglomeration of quantities and figures of little value for management purposes. Insist on detail if you pay for this service; specify the purpose for which the valuation is required to ensure that you get the required information.

F.I.F.O., L.I.F.O. AND AVERAGE COST

These headings refer to different bases for calculating cost of materials used in the productive process. They may be considered irrelevant to many small businesses and farms but it is helpful to understand the principles. They may well have relevance to intensive production systems and certainly it is well to be aware of the different methods of calculating cost and stock value.

F.I.F.O.

F.I.F.O. is an abbreviation for First In First Out: raw materials or goods are costed out in the same sequence as that in which they arrived in store. This method ensures that price increases are introduced into the productive process, or to sales, in the sequence in which they occur. On the other hand, it may be seen as an unnecessary and tedious chore in the case of a productive process which takes months to complete, such as bacon pig production where food is costed accurately. There are easier methods which provide the same degree of accuracy in such a situation.

The system is based on an 'In' record and an 'Out' record which includes date of delivery, description of goods, quantity, unit price and total price. Various methods may be devised to ensure correct sequence and pricing, but maintenance of a running total available and total requisitioned or allocated enables selection of the correct batch and price level. Store cards are available through business stationers which allow such a system, or a custom made layout to suit particular circumstances may be preferred.

L.I.F.O.

L.I.F.O. is simply a reversal of allocation procedure, whereby the last in is first out. This system may be chosen in a situation where it is considered important to reflect current prices in the cost of goods produced or sold.

Average cost

Average cost must be calculated on a **weighted average** basis in order to give an accurate unit cost of raw materials.

	£
1 tonne @£42/t	42
4 tonne @£52/t	208
5 tonne Total available	250 Total cost

$£\ \frac{250}{5} = £50$ per tonne

A simple average would be calculated by adding unit prices for each price level and dividing by the number of price levels used.

$£42 + £52 = £94 \div 2 = £47$ per tonne

and it can immediately be seen that this is inaccurate: $5 \times £47 = £235$!

CHAPTER 9

Bringing it all together

Now to bring the whole picture together we will produce a complete set of accounts, after completing the final stages of book-keeping for the year. It is essential that the procedure at each stage is fully understood, in order that the whole accounting process can be seen to fit together with mechanical precision. Each stage is as important to the next stage as are gears in a gearbox; one cog drives the next one but it must be of the right size, with no teeth missing!

To recap our progress to date, the accounts start and finish with a balance sheet. The book-keeping process in the intervening trading period allows us to produce the trading and profit and loss account. Providing all detail is correctly entered, a capital account can then be drawn up which also shows that the 'cogs' do indeed fit. A flow of funds statement then allows us to analyse the influence of transactions on our bank balance. All this gives us confidence that at least we know what has happened in the past and provides a basis on which to plan for the future.

The figures contained in the example accounts in this chapter may have become 'dated' by comparison to current prices, but this is of little importance. It is the procedure which is important and which can be applied to current information.

The business to which the accounts relate is a rented 110.2 ha mixed arable and dairy farm. (NB: The business is entirely fictional, and any similarity to an existing farm is purely coincidental.) We start by looking at the balance sheet (see next page) which was drawn up on the last day of the previous trading year, being the closing balance sheet for that year and the opening balance sheet for the trading year under consideration.

We have in this balance sheet a picture of the business as it existed at the beginning of the trading year. We shall examine the strengths and weaknesses of the balance sheet at a later stage but for the moment it may be noted that the owner's stake in the business, that is the equity, is reasonably substantial being well in excess of half of the total assets.

Green Farm Balance Sheet as at 31.3.X8

	£	£
Fixed assets		
Machinery W.D.V.	44,720	
Fixed equipment W.D.V.	912	
Dairy cows (90)	45,000	90,632
Current assets		
Dairy youngstock (38)	13,300	
Barley 4.25t	475	
Purchased concentrates	300	
*Seeds	3,210	
*Fertiliser	4,753	
*Crop protection	1,764	
Diesel and lubricating oil	600	24,402
Sundry debtors	9,867	9,867
		124,901

	£	£
Current liabilities		
Sundry creditors	7,443	
Bank overdraft	30,120	37,563
Long term liabilities		
Loan		7,614
Capital		79,724
		124,901

*In store and in the ground.

Year Ending 31. 3. X9

PAYMENTS

	Date	Detail	BANK		Cheque No.	CONTRA £	p.	DAIRY COW FEED £	p.	YOUNGSTOCK FEED £	p.	VET & MED £	p.	LIVESTOCK SUNDRIES £	p.	SEED £	p.	FERTILISER £	p.	CROP PROTECTION £	p.
1		APRIL b.f.	6177	92	-	81	24	1432	80	-		117	21	-		-		-		87	26
2		MAY b.f.	13008	61	-	34	96	1464	23	-		152	81	279		-		673		1462	-
3		JUNE b.f.	13571	03	-	34	96	1286	28	-		113	96	372	56	-		-		-	
4		JULY b.f.	6865	20	-	52	71	1468	25	-		149	83	118	-	-		-		-	
5		AUG b.f.	8358	79	-	48	14	477	12	-		195	30	-		-		-		-	
6		SEPT b.f.	16637	31	-	842	97	86	39	675		146	99	424	-	865	-	-		-	
7		OCT b.f.	16390	37	-	49	72	1190	61	148	-	334	57	218	-	1463	-	3244	-	1896	-
8		NOV b.f.	11409	79	-	214	41	1970	63	-		212	08	563	-	-		-		-	
9		DEC b.f.	13298	27	-	149	17	3541	86	152	-	264	07	640	44	651	-	-		422	-
10		JAN b.f.	10080	24	-	221	-	1620	76	200	-	471	12	516	-	247	-	2156	-	507	-
11		FEB b.f.	12368	57	-	129	10	3654	40	398		243	31	237	-	-	-	-	-	300	24

Year Ending 31. 3. X9

RECEIPTS

	Date	Detail	AMOUNT CHEQUES		Cheque No.	BANKED £	p.	CONTRA £	p.	MILK £	p.	CALVES £	p.	CULL COWS £	p.	BARREN HEIFERS £	p.	W. WHEAT HARVEST X8 £	p.	W. BARLEY HARVEST X8 £	p.
1		APRIL bf				7581	23	81	24	6192	91	563	56	330	-	-	-	-	-	-	-
2		MAY b.f.				10805	28	34	96	5607	54	-	-	-	-	-	-	-	-	-	-
3		JUNE b.f.				6175	78	34	96	5396	24	300	-	-	-	-	-	-	-	-	-
4		JULY b.f.				7026	38	52	71	6687	44	-	-	-	-	-	-	-	-	-	-
5		AUG b.f.				13500	25	48	14	5070	16	284	-	3840	-	-	-	-	-	-	-
6		SEPT b.f.				12353	79	842	97	5094	01	3820	44	3704	-	-	-	-	-	-	-
7		OCT b.f.				10432	17	49	72	5997	89	364	-	860	-	-	-	-	-	2480	-
8		NOV b.f.				15729	86	214	41	6451	34	620	-	1376	-	460	-	4280	-	1228	-
9		DEC b.f.				16962	48	149	17	6926	69	-	-	-	-	876	-	8946	-	-	-
10		JAN b.f				17748	36	221	-	8842	30	700	-	-	-	600	-	6640	-	-	-
11		FEB b.f.				16743	02	129	10	9541	62	775	-	-	-	-	-	5707		-	-

THE BOOK-KEEPING PROCEDURE

In order to demonstrate detail of the book-keeping but avoid unnecessary tedium, the first eleven months' work has been completed. Payments and receipts were recorded monthly, balanced and reconciled with the bank statement. Monthly column totals were then carried forward to a summary sheet shown below. (Figure 9.1)

To confirm that the work is balanced at this stage, follow checking procedures detailed on page 34.

The cash analysis

Full detail of the cash analysis for the final month in the business year is shown in Figure 9.2 on pages 76–77. It is up to the individual to determine just how much detail is included in the analysis, dependent upon the use to which it is to be put. In this case, detail is analysed even further in the gross margin data sheets in Chapter 10. The cash analysis in Figure 9.2 allows the development of a detailed profit and loss account.

Figure 9.1 **(below and facing page)** ***Summary balance sheets for 11 months***

Year Ending 31. 3. x9

PAYMENTS

	FUEL. OIL. ELECTRICITY £ p.	MACHINERY COSTS £ p.	CONTRACT £ p.	WAGES £ p.	GENERAL REPAIRS £ p.	RENT. RATES WATER £ p.	GEN. INS. SUNDRY F.C. £ p.	ADMIN. TELEPHONE. £ p.	LOAN REPAY. INTEREST. £ p.	CAPITAL MACHINERY £ p.	VAT on PURCH & to C+E £ p.	PRIVATE DRAWING £ p.	
1	91 34	410 -	148	1625 -	- -		–	141 78	–	820 -	225 77	1160 -	1
2	220 -	367 28	- -	1843 -	- -		2007 -	130 -	2710 75	- -	514 50	1220 -	2
3	383 21	1473 16	- -	1927 35	373 84	3618 80	367 18	-	- -	127 50	391 65	3170 50	3
4	- -	- -	558 -	1740 20	- -		294 -	538 40	- -	- -	730 23	1321 -	4
5	- -	493 23	-	1679 -	326 24		270 -	860	2691 63	- -	209 25	1205 16	5
6	1893 60	2041 -	-	2133 -	- -	3765 62	- -	657 68	-	- -	480 -	4312 -	6
7	167 30	326 15	-	1427 -	267 -		- -	163 18	-	2590 50	1341 30	1163 48	7
8	569 -	869 12	-	1807 -	133 34		363 -	874 -	2660 28	- -	375 75	1227 -	8
9	832 -	230 -	-	1563 14	542 16	2602 94	200 -	126 43		- -	329 40	1350 -	9
10	- -	658 20	-	1578 33	231 41		210 -	–	-	- -	687 -	1198 36	10
11	597 35	713 21	-	1506 83	190 39		260 82	120 04	2580 34	- -	478 23	1217 51	11

Year Ending 31. 3. x9

RECEIPTS

	W. OATS. HARVEST x8 £ p.	W. BARLEY HARVEST x7 £ p.	MISC. SALES. £ p.	CAPITAL MACHINERY £ p.	VAT from SALES + C.E. £ p.	PRIVATE & TRAN. IN. £ p.							
1	- -	- -	- -	- -	576 -	- -							1
2	- -	2850 -	156 93	- -	225 77	2000 -							2
3	- -	- -	- -	- -	514 50	- -							3
4	- -	- -	- -	- -	391 65	- -							4
5	- -	- -	624 -	- -	730 23	3000 -							5
6	- -	- -	79 06	- -	494 25	-							6
7	- -	- -	- -	300 -	480 -	- -							7
8	- -	- -	187 63	- -	1341 30	- -							8
9	- -	-	38 49	- -	324 47	- -							9
10	530	- -	327 66	- -	329 40	- -							10
11	- -	- -	161 50	- -	687 -	-							11

Figure 9.2 (below and facing page) *Cash analysis for the final month of the year*

PAYMENTS

Year Ending 31. 3. x9

	Date MARCH	Detail	BANK £ p.	Cheque No.	CONTRA £ p.	DAIRY COW FEED £ p.	YOUNGSTOCK FEED £ p.	VET + MED £ p.	LIVESTOCK SUNDRIES £ p.	SEED £ p.	FERTILISER £ p.	CROP PROTECTION £ p.
1	2	PRIVATE	530 -	0851 ✓								
2	2	SPAREPARTCO. Ltd	57 50	52 ✓								
3	5	MOBILE FEED SERVICE	84 17	53 ✓		84 17						
4	7	WAGES	785 -	54 ✓								
5	12	TELECOM. PHONE	101 40	55 ✓								
6	12	THE FORGE. GATE HINGES	22 93	56 ✓								
7	12	ELECTRA CO. CALCULATOR	40 96	57 ✓								
8	12	VETCO. WORMER	16 09	58				13 99				
9	16	DALGETY DAIRY CAKE	859 76	59 ✓		859 76						
10	16	JET STATION. PETROL	32 60	60 ✓								
11	17	C.W.G. Dairy Sund.	58 20	61 ✓				43 80				
12	17	DAVIES. PUMP REPAIRS	43 47	62 ✓				37 80				
13	18	PRIVATE	22 99	63								
14	18	AMOCO. LUB. OIL	105 -	65								
15	18	WILSON. D. CAKE. + COW CARDS	961 41	66 ✓		944 50						
16	21	BRADGATES. MINERALS	175 -	67 ✓		175 -						
17	22	CHEQUE CASHED	28 59	68 ✓								
18	24	BRADGATES. D. CAKE	1087 12	69 ✓		1087 12						
19	24	NFU. VEHICLE INS.	1062 62	70 ✓								
20	24	CRAIG + MILLER. VET	85 94	71				74 91				
21	24	PRIVATE	530 -	72 ✓								
22	24	JET STATION. PETROL	31 60	73								
23	27	RENT. (FURTHER FIELD)	213 55	74								
24	27	RENT. FLICHET + MOSS	4378 -	75								
25	28	WAGES.	813 15	76								
26	30	C.W.G.	1150 13	77				42 -	204 -		724 -	27 50
27	30	ATKINS MILL (FINAL SETTLEMENT)	1359 12	78		1359 12						
28	30	CONTRACT MAINT.	27 60	D.D								
29		MMB CONTRA	- -		68 62			62 25				
30		TOTAL C. Forward	14663 90		68 62	4509 67	—	274 75	204	—	724	27 50
31–38												
		TOTALS										

Bank + Contra: 14732·52

Dairy Cow Feed to Crop Protection: 14732·52

Year Ending 31. 3. x9

RECEIPTS

	Date MARCH	Detail	AMOUNT CHEQUES	Cheque No.	BANKED £ p.	CONTRA £ p.	MILK £ p.	CALVES £ p.	CULL COWS £ p.	BARREN HEIFERS £ p.	W. WHEAT HARVEST x8 £ p.	W. BARLEY HARVEST x8 £ p.
1	4	J. JOHNSON 1 B. Calf.	75 -					75 -				
2	4	WOLDS LIVERY	413 -		488 - ✓							
3	6	THELWELL. GRAZING. HAY	26 73		26 73 ✓							
4	16	MMB FEB MILK	9702 24		9702 24 ✓	68 62	9770 86					
5	21	C + E VAT	463 43		463 43 ✓							
6	28	ANGLIA GRAIN GROUP	3222 -		3222 -							3222 -
7		TOTAL C. Forward	13902 40		13902 40	68 62	9770 86	75 -	—	—	—	3222 -
8–15												

Banked + Contra: 13971·02

Milk to W. Barley: 13971·02

Year Ending 31.3.x9

PAYMENTS

	FUEL OIL. ELECTRICITY £	p.	MACHINERY COST £	p.	CONTRACT £	p.	WAGES £	p.	GENERAL REPAIRS £	p.	RENT. RATES WATER. £	p.	GEN. INS SUNDRY F.C. £	p.	ADMIN. TELEPHONE £	p.	LOAN REPAY INTEREST £	p.	CAPITAL MACHINERY £	p.	VAT on PURCH → To C+E £	p.	PRIVATE DRAWING £	p.
1																							530	–
2			50																		7	50		
3																								
4							785	–																
5															88	17					13	23		
6									19	95											2	48		
7															35	61					5	35		
8																					2	10		
9																								
10	32	60																						
11			6	81																	7	59		
12																					5	67		
13																							22	99
14	105	–																						
15															14	71					2	20		
16																								
17			15	22							11	09									2	28		
18																								
19			1062	62																				
20																					11	03		
21																							530	–
22	31	60																						
23											213	55												
24											4378	–												
25							813	15																
26									2	61											150	02		
27																								
28					24	–															3	60		
29																					6	37		
30	169	20	1134	65	24	–	1598	15	22	56	4602	64	–		138	49	–		–		219	92	1082	99
31																								
32																								
33																								
34																								
35																								
36																								
37																								
38																								
Totals																								

Year Ending 31.3.x9

RECEIPTS

	W. OATS HARVEST x8 £	p.	W. BARLEY HARVEST x7 £	p.	MISC. SALES £	p.	CAPITAL MACHINERY £	p.	VAT from SALES + C.E £	p.	PRIVATE + TRAN. IN £	p.	£	p.	£	p.	£	p.	£	p.	£	p.	£	p.
1																								
2	413																							
3					26	73																		
4																								
5									463	43														
6																								
7	413		–		26	73	–		463	43	–													
8																								
9																								
10																								
11																								
12																								
13																								
14																								
15																								

When all of the payments and receipts for the month have been entered they **must** be checked against the bank statement for the month to ensure the following points:

- Cheque amounts entered in the cash analysis agree with cheque amounts shown on the bank statement.
- All bankers' orders, direct debits, credit transfers, bank interest, bank charges and any other non-cheque payments or receipts shown on the bank statement are entered in the cash analysis.
- All unpresented cheques are recorded and totalled to allow calculation of the true bank balance at the month end.

If this procedure is not followed, the cash analysis is of little value and is likely to be discarded by the accountant who prepares final accounts. Cash books which do not reconcile with the bank statement will lead to erroneous accounts which cannot be balanced.

Bank statement

The bank statement is shown in Figure 9.3. Note the method of marking both the statement entries and the cash book entries in order to determine which cheques have not yet been presented.

Bank reconciliation

The reconciliation calculation is shown in Figure 9.4 on pages 80–81. It is important to check through every detail of the reconciliation procedure in order to establish full understanding – and reassurance that the work is correct!

The summary sheet completed

The final month has now been entered on to the summary sheet and column totals entered and cross checked. 'Bank' plus 'Contra' equals the sum total of the analysis columns.

It is sometimes helpful to make a final check on the bank position by the following method:

- Calculate annual cash flow by taking total receipts and total payments from the summary sheet.
- Bank balance at the start of the year together with cash flow for the year gives the balance at the year end, which must be identical to the balance calculated for the final month. (Figure 9.5, pages 80–81)

	£
Bank at 31.3.88	30,120–
Total payments	142,830–
Total receipts	148,961+
Bank at 31.3.89	23,989–

(text continues on page 82)

Figure 9.3 Bank statement

ANYBANK — GREEN FARM ACCOUNT

DETAILS		PAYMENTS	RECEIPTS	DATE	BALANCE
				19X9	
BALANCE FORWARD				28FEB	22563.80DR
	000851√	530.00		1MAR	23093.80DR
	000852√	57.50		1MAR	23151.30DR
COUNTER CREDIT			488.00√	8MAR	
	000836√	92.23		8MAR	
	000850√	340.00		8MAR	23095.53DR
	000854√	785.00		9MAR	
COUNTER CREDIT			26.73√	9MAR	23853.80DR
	000853√	84.17		10MAR	
	000855√	101.40		10MAR	24039.37DR
	000833√	46.97		11MAR	
	000822√	184.50		11MAR	24270.84DR
	000856√	22.93		14MAR	
	000860√	32.60		14MAR	24326.37DR
MILK MARKETING BOARD	BGC		9702.24√	16MAR	
	000859√	859.76		16MAR	15483.89DR
HM CUSTOMS VAT	BGC		463.43√	21MAR	
	000867√	175.00		21MAR	15195.46DR
	000868√	28.59		24MAR	
	000872√	530.00		24MAR	
	000857√	40.96		24MAR	15795.01DR
	000861√	58.20		25MAR	
	000866√	961.41		25MAR	16814.62DR
	000869√	1087.12		28MAR	17901.74DR
	000870√	1062.62		29MAR	
	000862√	43.47		29MAR	
DIRECT DEBIT	DDR√	27.60		29MAR	19035.43DR

ABBREVIATIONS: DIV Dividend STO Standing Order BGC Bank Giro Credit DDR Direct Debit DR Overdrawn Balances

Figure 9.4 (right and facing page)
Bank reconciliation

February 19X9 Unpresented cheques

Payments No.	£	Receipts
822	184.50	—
833	46.97	
836	92.23	
850	340.00	
	663.70	

Bank statement	£	*Cash Analysis February 19X9*	£
Balance at 28.2.X9	22,563.80–	Balance b.f. Jan	27,601.95–
Unpresented payments	663.70–	Payments	12,368.57–
Unpresented receipts	–	Receipts	16,743.02+
	23,227.50–		23,227.50–

Figure 9.5 (below and facing page)
Summary sheets for 12 months

Year Ending 31. 3. X9

PAYMENTS

	Date	Detail	BANK		Cheque No.	CONTRA £	p.	DAIRY COW FEED £	p.	YOUNGSTOCK FEED £	p.	VET & MED £	p.	LIVESTOCK SUNDRIES £	p.	SEED £	p.	FERTILISER £	p.	CROP PROTECTION £	p.
1		APRIL b.f.	6177	92	–	81	24	1432	80	–		117	21	–		–		–		87	26
2		MAY b.f.	13008	61	–	34	96	1464	23	–		152	81	279		–		673		1462	–
3		JUNE b.f.	13571	03	-	34	96	1286	28	–		113	96	372	56	-		–		–	
4		JULY b.f.	6865	20	–	52	71	1468	25	-		149	83	118	–	–		–		–	
5		AUG b.f.	8358	79	–	48	14	477	12	–		195	30	–		–		–		–	
6		SEPT b.f.	16637	31	-	842	97	86	39	675		146	99	424	–	865	–	–		–	
7		OCT b.f.	16390	37	-	49	72	1190	61	148	-	334	57	- 218	-	1463	–	3244	-	1896	-
8		NOV b.f.	11409	79	–	214	41	1970	63	–		212	08	563	-	–		–		–	
9		DEC b.f.	13298	27	–	149	17	3541	86	152	-	264	07	640	44	651	-	-		422	-
10		JAN b.f.	10080	24	–	221	-	.620	76	200	–	471	12	516	-	247	-	2156	-	507	–
11		FEB b.f.	12368	57	–	129	10	3654	40	398		243	31	237	-	–	-	–	-	300	24
12		MAR b.f.	14663	90		68	62	4509	67	–		274	75	204		–		724	-	27	50
13		TOTAL PAYMENTS	142830	-		1927	-	22703	-	1573		2676	-	3572	-	3226	-	6797	-	4702	–

Year Ending 31. 3. X9

RECEIPTS

	Date	Detail	AMOUNT CHEQUES		Cheque No.	BANKED £	p.	CONTRA £	p.	MILK £	p.	CALVES £	p.	CULL COWS £	p.	BARREN HEIFERS £	p.	W. WHEAT HARVEST X8 £	p.	W. BARLEY HARVEST X8 £	p.
1		APRIL bf				7581	23	81	24	6192	91	563	56	330	-	-	-	-	-	-	-
2		MAY b.f.				10805	28	34	96	5607	54	-	-	-	-	-	-	-	-	-	-
3		JUNE b.f.				6175	78	34	96	5396	24	300	-	-	-	-	-	-	-	-	-
4		JULY b.f.				7026	38	52	71	6687	44	-	-	-	-	-	-	-	-	-	-
5		AUG b.f.				13500	25	48	14	5070	16	284	-	3840	-	-	-	-	-	-	-
6		SEPT b.f.				12353	79	842	97	5094	01	3820	44	3704	-	-	-	-	-	-	-
7		OCT b.f.				10432	17	49	72	5997	89	364	-	860	-	-	-	-	-	2480	-
8		NOV b.f.				15729	86	214	41	6451	34	620	-	1376	-	460	-	4280	-	1228	-
9		DEC b.f.				16962	48	149	17	6926	69	-	-	-	-	876	-	8946	-	-	-
10		JAN b.f.				17748	36	221	-	8842	30	700	-	-	-	600	-	6640	-	-	-
11		FEB b.f.				16743	02	129	10	9541	62	775	-	-	-	-	-	5707		-	-
12		MAR b.f.				13902	40	68	62	9770	86	75	-	-	-	-	-	-		3222	-
13		TOTAL RECEIPTS				148961	–	1927	–	81579	–	7502	–	10110	–	1936	–	25573		6930	-

March 19X9 Unpresented cheques

Payments No.	£	*Receipts* £
58	16.09	3,222
63	22.99	
65	105.00	
71	85.94	
73	31.60	
74	213.55	
75	4,378.00	
76	813.15	
77	1,150.13	
78	1,359.12	
	8,175.57	

Bank statement	£	*Cash Analysis March 19X9*	£
Balance at 31.3.X9	19,035.43−	Balance b.f. Feb	23,227.50−
Unpresented payments	8,175.57−	Payments	14,663.90−
Unpresented receipts	3,222.00+	Receipts	13,902.40+
	23,989.00−		23,989.00−

Year Ending 31. 3. X9

PAYMENTS

	FUEL. OIL. ELECTRICITY £	p.	MACHINERY COSTS £	p.	CONTRACT £	p.	WAGES £	p.	GENERAL REPAIRS £	p.	RENT. RATES WATER £	p.
1	91	34	410	-	148	-	1625	-	—	-		
2	220	-	367	28	-	-	1843	-	-	-		
3	383	21	1473	16	-	-	1927	35	373	84	3618	80
4	-	-	-	-	558	-	1740	20	-	-		
5	-	-	493	23	-		1679	-	326	24		
6	1893	60	2041	-	-		2133	-	-	-	3765	62
7	167	30	326	15	-		1927	-	267	-		
8	569	-	869	12	-		1807	-	133	34		
9	832	-	230	-	-		1563	14	542	16	2602	94
10	-	-	658	20	-		1598	33	231	47		
11	597	35	713	21	-		1506	83	190	39		
12	169	20	1134	65	24		1598	15	22	56	4602	64
13	4923	-	8716	-	730		20948	-	2087	-	14590	-

	GEN. INS. SUNDRY F.C. £	p.	ADMIN. TELEPHONE. £	p.	LOAN REPAY. INTEREST. £	p.	CAPITAL MACHINERY £	p.	VAT on PURCH + to C+E £	p.	PRIVATE DRAWING £	p.
1	—		141	78	—		820	-	225	77	1160	-
2	2007	-	130	-	2710	75	-	-	514	50	1220	-
3	367	18	-	-	-	-	127	50	391	65	3170	50
4	294	-	538	40	-	-	-	-	730	23	1321	-
5	270	-	860		2691	63	-	-	209	25	1205	16
6	-		657	68	-		-	-	480	-	4312	-
7	-	-	163	18	-		2590	50	1341	30	1163	48
8	363	-	874	-	2660	28	-	-	375	75	1227	-
9	200	-	126	43	-		-	-	329	40	1350	-
10	210	-	-		-		-	-	687	-	1198	36
11	260	82	120	04	2580	34	-	-	478	23	1217	51
12	-	-	138	49	-		-	-	219	92	1082	99
13	3912		3750	-	10643	-	3538	-	5983	-	19628	-

Year Ending 31. 3. X9

RECEIPTS

	W. OATS. HARVEST X8 £	p.	W. BARLEY HARVEST X7 £	p.	MISC. SALES. £	p.	CAPITAL MACHINERY £	p.	VAT from SALES + C.E. £	p.	PRIVATE & TRAN. IN. £	p.
1	—	-	-	-	—	-	—	-	576	-	—	-
2	-	-	2850	-	156	93	-	-	225	77	2000	-
3	-	-	-	-	-	-	-	-	514	50	—	-
4	-	-	-	-	-	-	-	-	391	65	-	-
5	-	-	-	-	624	-	-	-	730	23	3000	-
6	-	-	-	-	79	06	-	-	499	25	—	-
7	-	-	-	-	-	-	300	-	480		-	-
8	-	-	-	-	187	63	-	-	1341	30	-	-
9	-	-	-	-	38	49	-	-	324	47	-	-
10	530	-	-	-	327	66	-		329	40	-	-
11	-	-	-	-	161	50	-	-	687	-	—	-
12	413	-	-	-	26	13	-	-	463	43	-	-
13	943	-	2850	-	1602	-	300	-	6563	-	5000	-

Adjustments to the summary sheet

As indicated in Chapter 4, the cash analysis is a record of payments and receipts made during the year, it is not a record of transactions made by the business during the accounting year. The payments and receipts made during the year include settlement of debts owed (in both directions) at the beginning of the year but since they belong to the previous accounting year they must be deducted from the column totals.

At the end of the year, it is seen that there are unpaid bills for transactions entered into during the accounting year. (Figure 9.6) These must be added to the column totals.

The final adjusted column totals now represent transactions for the year, adjusted payments being **expenditure** and adjusted receipts being **revenue**. (Figure 9.7, pages 84–85)

VALUATION FOR THE YEAR

Valuation method has been dealt with in Chapter 8 so detail only is provided for this set of accounts.

Schedule of Valuations at 31.3.X9

Livestock	*Quantity*	£	£
Dairy cows	78	39,000	
Youngstock	43	13,700	52,700
Crops and stores			
Barley homegrown	1 tonne	105	
Concentrate feed	11.5 tonnes	1,650	
Seeds		3,226	
Fertiliser		5,540	
Crop protection		1,170	
Fuel and oil		550	12,241
	Total		64,941

THE PROFIT AND LOSS ACCOUNT

Now we move from the mere 'chores' of the accounting process to the more exciting stages. We bring together the influence of valuation change, revenue, expenditure and notional costs to determine profit for the trading period.

It is important that adjusted cash analysis column totals are marked with the destination to which they are carried forward – the profit and loss account, the depreciation account or the capital account – in order that all totals can be seen to have been correctly transferred. It can be most annoying to find that the accounts do not balance due to a cash analysis total hiding away in a corner, not taken forward to its proper place in the accounts. (Figure 9.8, page 86)

(text continues on page 87)

Figure 9.6 Sundry debtors and creditors

At 31.3.X8

Debtors		Total invoice	VAT	Invoice excluding VAT
		£	£	£
M.M.B.	Milk	6,193		
J. Pick	Calves 3	273		
D. Griffin	Heifer	450		
Drapers	Barley	2,375		
C.&E.	VAT claim	576		
		9,867		

Creditors		Total invoice	VAT	Invoice excluding VAT
		£	£	£
Drapers	Fertiliser	3,278	427	2,851
Dalgety	Dairy feed	2,950		2,950
John Bull	Insurance broker	430		430
N.R. Garage	Diesel fuel	650		650
Vetco	Vet	135	17	118
		7,443	444	6,999

At 31.3.X9

Debtors		Total invoice	VAT	Invoice excluding VAT
		£	£	£
M.M.B.	March milk	9,674		
Meatco	Cull cow	430		
J. Sope	Calves 2	186		
Millfeeds Ltd	Wheat	2,740		
T.M.S.	Sprayer	230	30	200
C.&E.	VAT claim	411		
		13,671	30	200

Creditors		Total invoice	VAT	Invoice excluding VAT
		£	£	£
Drapers	Fertiliser	5,437	709	4,728
Dalgety	Dairy cake	1,850		1,850
Vetco	Vet	310	40	270
Bildit & Co	Repairs	753	98	655
N.R. Garage	Petrol	187	24	163
M.M.B. (contra)	Dairy sundry	138	18	120
Brewers	Youngstock feed	120		120
		8,795	889	7,906

Figure 9.7** (below and facing page) **Adjusted summary sheet

Year Ending 31. 3. X9

PAYMENTS

	Date	Detail	BANK		Cheque No.	CONTRA		DAIRY COW FEED		YOUNGSTOCK FEED		VET & MED		LIVESTOCK SUNDRIES		SEED		FERTILISER		CROP PROTECTION	
						£	p.	£	p.	£	p.	£	p.	£	p.	£	p.	£	p.	£	p.
1		APRIL b.f.	6177	92	–	81	24	1432	80	–		117	21	–		–		–		87	26
2		MAY b.f.	13008	61	–	34	96	1464	23	–		152	81	279		–		673		1462	–
3		JUNE b.f.	13571	03	–	34	96	1286	28	–		113	96	372	56	–		–		–	
4		JULY b.f.	6865	20	–	52	71	1468	25	–		149	83	118	–	–		–		–	
5		AUG b.f.	8358	79	–	48	14	477	12	–		195	30	–		–		–		–	
6		SEPT b.f.	16637	31	–	842	97	85	39	675		146	99	424	–	865	–	–		–	
7		OCT b.f.	16390	37	–	49	72	1190	61	148	–	334	57	218	–	1463	–	3244	–	1896	–
8		NOV b.f.	11409	79	–	214	41	1970	63	–		212	08	563	–	–		–		–	
9		DEC b.f.	13298	27	–	149	17	3541	86	152	–	264	07	640	44	651	–	–		422	–
10		JAN b.f.	10080	24	–	221	–	1620	76	200	–	471	12	516	–	247	–	2156	–	507	–
11		FEB b.f.	12368	57	–	129	10	3654	40	398		243	31	237	–	–	–	–	–	300	24
12		MAR b.f.	14663	90		68	62	4509	67	–		274	75	204		–		724	–	27	50
13		TOTAL PAYMENTS	142830	–		1927	–	22703	–	1573		2676	–	3572	–	3226	–	6797	–	4702	–
14		Less o. creditors						2950	–	–		118		–		–		2851		–	
15		–						19753		–		2558		–		–		3946		–	
16		Plus c. creditors						1850		120		270		120		–		4728		–	
17		EXPENDITURE						21603		1693		2828		3692		3226		8674		4702	
18																					
19								P+L		P+L		P+L		P+L		P+L		P+L		P+L	
20																					
21																					
22																					
23																					
24																					
25																					
26																					
27																					

Year Ending 31. 3. X9

RECEIPTS

	Date	Detail	AMOUNT CHEQUES		Cheque No.	BANKED		CONTRA		MILK		CALVES		CULL COWS		BARREN HEIFERS		W. WHEAT HARVEST x8		W. BARLEY HARVEST x8	
						£	p.	£	p.	£	p.	£	p.	£	p.	£	p.	£	p.	£	p.
1		APRIL bf				7581	23	81	24	6192	91	563	56	330	–	–	–	–	–	–	–
2		MAY b.f.				10805	28	34	96	5607	54	–	–	–	–	–	–	–	–	–	–
3		JUNE b.f.				6175	78	34	96	5396	24	300	–	–	–	–	–	–	–	–	–
4		JULY b.f.				7026	38	52	71	6687	44	–	–	–	–	–	–	–	–	–	–
5		AUG b.f.				13500	25	48	14	5070	16	284	–	3840	–	–	–	–	–	–	–
6		SEPT b.f.				12353	79	842	97	5094	01	3820	44	3704	–	–	–	–	–	–	–
7		OCT b.f.				10432	17	49	72	5997	89	364	–	860	–	–	–	–	–	2480	–
8		NOV b.f.				15729	86	214	41	6451	34	620	–	1376	–	460	–	4280	–	1228	–
9		DEC b.f.				16962	48	149	17	6926	69	–	–	–	–	876	–	8946	–	–	–
10		JAN b.f.				17748	36	221	–	8842	30	700	–	–	–	600	–	6640	–	–	–
11		FEB b.f.				16743	02	129	10	9541	62	775	–	–	–	–	–	5707		–	–
12		MAR b.f.				13902	40	68	62	9770	86	75	–	–	–	–	–	–		3222	–
13		TOTAL RECEIPTS				148961	–	1927	–	81579	–	7502	–	10110	–	1936	–	25573		6930	–
14		LESS o.debtors								6193	–	273	–	–		450	–	–		–	
15										75386	–	7229	–	–		1486	–	–		–	
16		Plus c.debtors								9674	–	186	–	430	–	–		2740	–	–	
17		REVENUE								85060	–	7415	–	10540	–	1486	–	28313		6930	
18																					
19										P+L		P+L		P+L		P+L		P+L		P+L	
20																					
21																					
22																					

Year Ending 31. 3. x9

PAYMENTS

FUEL. OIL. ELECTRICITY		MACHINERY COSTS		CONTRACT		WAGES		GENERAL REPAIRS		RENT. RATES WATER		GEN. INS. SUNDRY F.C.		ADMIN. TELEPHONE.		LOAN REPAY. INTEREST.		CAPITAL MACHINERY		VAT on PURCH + to C+E		PRIVATE DRAWING		
£	p.	£	p.	£	p.	£	p.	£	p.	£	p.	£	p.	£	p.	£	p.	£	p.	£	p.	£	p.	
91	34	410	-	148	-	1625	-	-	-			-		141	78	-		820	-	225	77	1160	-	1
220	-	367	28	-	-	1843	-	-	-			2007	-	130	-	2710	75	-	-	514	50	1220	-	2
383	21	1473	16	-	-	1927	35	373	84	3618	80	367	18	-	-	-	-	127	50	391	65	3170	50	3
-	-	-	-	558	-	1740	20	-	-			294	-	538	40	-	-	-	-	730	23	1321	-	4
-	-	493	23	-		1679	-	326	24			270	-	860		2691	63	-	-	209	25	1205	16	5
1893	60	2041	-	-		2133	-	-	-	3765	62	-	-	657	68	-		-	-	480	-	4312	-	6
167	30	326	15	-		1927	-	267	-			-	-	163	18	-		2590	50	1341	30	1163	48	7
569	-	869	12	-		1807	-	133	34			363	-	874	-	2660	28	-	-	375	75	1227	-	8
832	-	230	-	-		1563	14	542	16	2602	94	200	-	126	43	-		-	-	329	40	1350	-	9
-	-	658	20	-		1598	33	231	47			210	-	-		-		-	-	687	-	1198	36	10
597	35	713	21	-		1506	83	190	34			260	82	120	04	2580	34	-	-	478	23	1217	51	11
169	20	1134	65	24		1598	15	22	56	4602	64	-	-	138	49	-		-	-	219	92	1082	99	12
4923	-	8716	-	730		20948	-	2087	-	14590	-	3972		3750	-	10643	-	3538	-	5983	-	19628	-	13
650	-	-		-		-		-		-		430	-	-		-		-		444		-		14
4273	-	-		-		-		-		-		3542	-	-		-		-		5539		-		15
163	-	-		-		-		655	-	-		-		-		-		-		889		-		16
4436	-	8716	-	730	-	20948	-	2742	-	14590	-	3542	-	3750		10643		3538		6428		19628		17
																								18
P+L		P+L		P+L		P+L		P+L		P+L		P+L		P+L				Dep		VAT		Cap		19
																		A/C		A/C		A/C		20
																								21
																								22
														Repayment		5000		to Cap. A/c						23
														Interest		5643		to P+L A/c						24
																10643								25
																								26
																								27

Year Ending 31. 3. x9

RECEIPTS

W. OATS. HARVEST x8		W. BARLEY HARVEST x7		MISC. SALES.		CAPITAL MACHINERY		VAT from SALES + C.E.		PRIVATE + TRAN. IN.		
£	p.	£	p.	£	p.	£	p.	£	p.	£	p.	
-	-	-	-	-	-	-	-	576	-	-	-	1
-	-	2850	-	156	93	-	-	225	77	2000	-	2
-	-	-	-	-	-	-	-	514	50	-	-	3
-	-	-	-	-	-	-	-	391	65	-	-	4
-	-	-	-	624	-	-	-	730	23	3000	-	5
-	-	-	-	79	06	-	-	494	25	-	-	6
-	-	-	-	-	-	300	-	480	-	-	-	7
-	-	-	-	187	63	-	-	1341	30	-	-	8
-	-	-	-	38	49	-	-	324	47	-	-	9
530	-	-	-	327	66	-	-	324	40	-	-	10
-	-	-	-	161	50	-	-	687	-	-	-	11
413	-	-	-	26	73	-	-	463	43	-	-	12
943	-	2850	-	1602	-	300	-	6563	-	5000	-	13
-	-	2375	-	-	-	-		576	-	-	-	14
-	-	475	-	-	-	-		5987	-	-	-	15
-	-	-		-	-	200	-	441	-	-	-	16
943	-	475		1602	-	500	-	6428	-	5000	-	17
												18
P+L		P+L		P+L		Dep		VAT		Cap		19
						A/c		A/c		A/c		20
												21
												22

Figure 9.8 Green Farm profit and loss account for year ending 31.3.X9

Valuation at 1.4.X8	£	£	*Valuation at 31.3.X9*	£	£
Dairy cows	45,000		Dairy cows	39,000	
Youngstock	13,300		Youngstock	13,700	
Barley (homegrown)	475		Barley (homegrown)	105	
Concentrate feed	300		Concentrate feed	1,650	
Seeds	3,210		Seeds	3,226	
Fertiliser	4,753		Fertilser	5,540	
Fuel	600		Fuel	550	
Crop protection	1,764	69,402	Crop protection	1,170	64,941
Expenditure			*Revenue*		
Dairy cow feed	21,603		Milk	85,060	
Youngstock feed	1,693		Calves	7,415	
Vet. and med.	2,828		Cull cows	10,540	
Livestock sundries	3,692		Barren heifers	1,486	
Seed	3,226		Winter Wheat		
Fertiliser	8,674		(Harvest year × 8)	28,313	
Crop protection	4,702		Winter Barley		
Fuel/oil/electricity	4,436		(Harvest year × 8)	6,930	
Machinery costs	8,716		Winter Oats		
Contract	730		(Harvest year × 8)	943	
Wages	20,948		Winter Barley		
General repairs	2,742		(Harvest year × 7)	475	
Rent/rates/water	14,590		Miscellaneous	1,602	142,764
General insurance/sundry sundry fixed costs	3,542				
Administration/telephone	3,750		*Non-cash receipts*		
Bank interest and charges	5,643	111,515	Heifer – to house	450	
			Milk – to house	263	713
Depreciation					
Machinery	9,552				
Fixed equipment	152	9,704			
		190,621			
Profit		17,797			
		208,418			208,418

CALCULATION OF DEPRECIATION

Machinery account

	£
W.D.V. at 31.3.X8	44,720
Purchases	3,538
	48,258
Sales	500
	47,758
Depreciation at 20% p.a.	9,552
W.D.V. at 31.3.X9	38,206

Fixed equipment account

	£
W.D.V. at 31.3.X8	912
Depreciate 10% cost £1,520	152
W.D.V. at 31.3.X9	760

THE BALANCE SHEET

Valuation detail, debtors, bank balance, creditors and loans are now compiled to produce the real pointer to the health of the business, the balance sheet. Consistency of style and layout is important in order to allow comparison with previous balance sheets and calculation of trends – are we going up or down? More of that in a later chapter.

Green Farm Balance sheet as at 31.3.X9

Fixed assets	£	£	*Current liabilities*	£	£
Machinery W.D.V.	38,206		Sundry creditors	8,795	
Fixed equipment W.D.V.	760		Bank overdraft	23,989	32,784
Dairy cows (78)	39,000	77,966			
			Long term liabilities		
Current assets			Loan		2,614
Dairy youngstock	13,700				
Barley – homegrown 1 t	105				
Purchased concentrates	1,650		Capital		82,180
*Seeds	3,226				
*Fertiliser	5,540				
*Crop protection	1,170				
Diesel and lubricating oil	550	25,941			
Sundry debtors	13,671	13,671			
		117,578			117,578

* In store and in the ground

THE CAPITAL ACCOUNT

Here we create a picture of what has happened to the equity, or capital, as a result of activities during the trading period. We take the capital at the beginning of the trading period as our starting point and add profit and any non-business funds brought into the business. Against this we set any withdrawal of funds for non-business purposes and of course loss, when this occurs.

Green Farm Capital Account for year ending 31.3.X9

	£		£
Loss for the year	–	Capital at 31.3.X8	79,724
Drawings	20,341	Profit for the year	17,797
Capital at 31.3.X9	82,180	Funds transferred in	5,000
	102,521		102,521

The equity, or capital shown in the capital account will always be in agreement with capital shown in the balance sheet for the same date – unless figures or calculations are erroneous and of course the errors must be traced and corrected. The layman often refers to the capital account as a 'proof' but this is a narrow viewpoint since the capital account should be seen as an informative document, not merely a check on the accuracy of the other documents in the accounting process.

THE FLOW OF FUNDS STATEMENT

This document simply sets out the factors which cause increase or decrease in the bank balance.

There has been many a puzzled look at the bank statement after receiving the accounts for a year of profitable trading: "It's all double Dutch to me. I make a decent profit and I'm even deeper in the red at the bank than I was a year ago". Well, help has arrived! The flow of funds statement shows clearly, in detail, the reasons for change in the bank position.

FLOW OF FUNDS – EXPLANATION

A flow of funds statement analyses the influences on bank balance, giving each positive and negative effect in detail. The starting point is the 'norm', or normal situation, against which the end situation – end of year, end of trading period or in fact any point at which accounts are produced – is compared. The **change** in bank balance is analysed (change in bank represents cash flow).

Green Farm Flow of Funds Statement for year ending 31.3.X9

Source of funds	£
Profit	17,797
Depreciation	9,704
Capital sales	500
Decrease in valuation	4,461
Decrease in debtors	–
Increase in creditors	1,352
Introduction of funds	5,000
Total A	38,814

Disposition of funds	
Loss	–
Capital purchases	3,538
Increase in valuation	–
Increase in debtors	3,804
Decrease in creditors	–
Loan repayment	5,000
Drawings	20,341
Total B	32,683
Flow of funds (A–B)	6,131

Change in bank position (cash flow or flow of funds)

	£
Bank at 31.3.X8	(30,120)
Bank at 31.3.X9*	(23,989)
Flow of funds	6,131

*The overdraft has decreased at the end of the year because of a positive flow of funds.

Check the figures with the help of notes given below.

Factors affecting cash flow

Profit In simplistic terms, profit may be expected to add to bank funds but other factors must be considered.

Depreciation is the estimated fall in value of capital assets such as machinery, but it is not an expense which results in payment of money from the bank. Depreciation included in the profit and loss account must be added back, increasing the influence of profit.

Capital purchases and sales Purchase and sale of capital items such as machinery do not feature in the profit and loss account since the annual fall in value is shown as depreciation. Purchase of capital items will cause an outflow or reduction of funds, sale of capital items will cause an inflow or addition to funds.

Valuation change If valuation of stock remains the same at each year end, the same amount of money is therefore invested in goods rather than being liquid in the bank. If valuations increase, extra money has been invested and there is a reduction, or outflow, of bank funds. If valuations decrease, some of the stock held has been liquified, or turned to cash, with a resultant addition to, or inflow of, bank funds.

Debtors If there is an increase in debts owed to the business at the year end, a change from the normal has occurred and additional funds are held by customers. The additional funds held by customers have a negative effect on bank balance. If there is a decrease in debts owed to the business, then money held by customers has been released and will increase money held in the bank.

Creditors A similar reasoning applies to that for debtors but in reverse. Money owed to creditors is effectively money held by the business; if there is an increased amount owed then that additional amount increases bank funds. If the amount owed to creditors decreases, debts are being paid earlier than normal and so money held but owing to creditors is reduced, with a consequent reduction of bank funds.

Introduction or withdrawal of funds New investment of funds will increase the bank balance and withdrawal of funds will have the opposite effect.

CHAPTER 10

What can the accounts tell us now?

GROSS MARGINS

An accurately prepared profit and loss account and balance sheet are hives of information – but not without more work! A traditionally prepared profit and loss account provides overall information on profit for the business; but where there is more than one source of profit from business activity then closer examination is required. It is possible that the good results of one enterprise are masking the failure of another.

The problem with a 'normal' profit and loss account is that factors influencing profit or loss are hidden in the valuations and in the fact that costs and output have not been calculated. Accounts for a farm with livestock enterprises are likely to include detail of opening and closing stock of feed, plus feed purchased during the year, but further calculations are required to determine cost of feed. Further examples of obscurity could be given but they will become obvious as the work in this chapter progresses.

Preparation of enterprise gross margins and a gross margin account should be considered a normal development of year end accounts. Records and accounts should be so maintained and prepared that gross margin calculation is simply seen as a part of year-end accounting.

What exactly are gross margins anyway? The term used by non-agricultural business is **contribution** rather than gross margin and it is perhaps more descriptive. Understanding the purpose of gross margins, or contribution, is dependent on a broad understanding of business and the accounting process in order that all parts may be seen in context.

Gross margins are a measurement of enterprise performance in isolation from (shared) overhead business costs. Output of an enterprise less the directly applicable (variable) costs determines gross margin. Gross margin is the contribution made towards meeting overhead costs, and of course profit, and so the success or otherwise of enterprises may be compared.

Comparison of gross margins for successive years may also determine trends to improved or declining results. Comparison of margins with other farms is also useful, but less reliable since there are so many background variables to consider – does soil type and fertility differ, are there different pressures and objectives for the other farms? There are many arguments to support or criticise gross margin analysis but a reasonable assessment of business performance can be made on this basis.

RECORDS FOR PRODUCING MARGINS

Essential to accurate enterprise analysis is the well thought out design and disciplined maintenance of financial and physical records. Planning well in advance and creating suitable recording systems – the simpler the better, so long as they are effective – will ensure that the required information is to hand when it is required. All too often there is a 'head in the sand' approach to recording and then production of gross margin information becomes a time consuming, frustrating affair with end results lacking the desired level of accuracy.

All of the work covered in earlier chapters is relevant to the present consideration, including the absolute necessity to reconcile financial records back to the bank and physical records back to physical stock checks.

Cash analysis

Analysis column headings must be carefully thought out to ensure that, within the space allowed, there is clear division of recorded information. Analysis of sales to relevant enterprises is usually easier than that for purchases in the case of a farm business. Where purchase of, e.g., fertiliser is recorded then it is likely that an additional supporting recording system will be necessary since the fertiliser may actually bc used for a number of enterprises.

The cash analysis book should preferably be balanced, ruled off and carried to the summary page on a monthly basis. It is important that all variable costs are allocated and entered into relevant gross margin records at month end as well. If it is left 'until there is time', it will not be done until year end and by that time vital detail of further breakdown may have been forgotten. Notes, maybe in pencil, in the cashbook act as an *aide mémoire* to help with allocation. Be inventive, but above all be consistent and 'strike while the iron is hot' – while the information is still fresh in your mind.

Double entry book-keeping

Points made about cash analysis of purchase and sales records apply equally to double entry book-keeping. Give careful consideration to break-down and enter both financial and quantitative detail.

Additional records

The type of business and individual enterprises will determine the additional records required to make a realistic and informative analysis. The areas for consideration are quantity (number/weight/volume), allocation and internal transfer (between enterprises). In many instances it is possible to include all of these areas in one record. Basic principles apply regardless of the type of business, be it producing tomatoes, garden sheds or pigs.

Quantity – number/weight/volume

The requirement to determine quantity may be that of average numbers, quantity produced or quantity used. In each case certain basic principles apply and it is essential that these principles are thoroughly understood.

Average numbers may be required for a variety of purposes, dependent on the enterprise. A stock check or head count should be made at regular intervals, perhaps monthly, and recorded. A record should also be maintained of additions to and deductions from opening stock – the balance taken at any time should agree with a head count/stock check. When additions and deductions are simple, such as raw material into and out of store, then all that is needed is a two column record.

	Date	Debit (or Received)	Credit (or Allocated)
Opening stock @	XX	500	
Purchased, delivered to store	XX	1,000	
Issued to A unit	XX		200
Issued to B unit	XX		300
Balance, being stock c.d. @	XX		1,000
		1,500	1,500
Opening stock (new period) b.d. @	XX	1,000	

There are situations when it is necessary to record the different sources and destinations of additions and deductions. It is necessary to record this detail for a livestock breeding enterprise (example given below) but it may equally apply to an intensive horticultural enterprise or other productive/commercial unit.

Livestock numbers for dairy herd

Month ending ....

	Cows in milk	Dry cows	Heifers over 2 yrs	Young stock 0–2 yrs	Bulls	Total	Notes
1 Opening stock							
2 Purchased							
3 Bred							
4 Transferred in							
Total A (1–4)							
5 Sold							
6 Died							
7 Transferred out							
Total B (5–7)							
Closing stock A–B							

Note: When transfer between classes of stock occur, transfer out of one class must be balanced by transfer in to another.

Quantity produced Here again it is necessary to use logic and knowledge of the enterprise together with calculations and when there is 'work in progress' at the start of an accounting period then it must be taken into consideration. Work in progress could include partially completed manufactured goods, farm livestock and possibly certain types of horticultural or arboricultural growing stock. Any value (quantitative) of stock at the beginning of a period cannot be considered to have been produced during the period so must be subtracted.

The calculation will include the following steps:

	No./wt./vol./£
Closing stock plus	
Sales	
Transfers out	____
Total A	____
Less	
Opening stock	
Purchases	
Transfers in	____
Total B	____
Quantity produced A–B	____

When the enterprise has a clear-cut production cycle, such as agricultural cash crops, then the opening stock is not part of the calculation. For example, grain in store from a previous harvest year is not taken into consideration when calculating yield for the current year.

Closing stock plus	
Sales	
Transfers out	
	———
Quantity produced	
	———

Specific productive or commercial enterprises may require slight variations on the above methods.

Quantity used To calculate the amount used, the calculation of amount produced is turned on its head.

	No./wt./vol./£
Opening stock plus	
Purchases	
Transfers in	
	———
Total available A	
	———
Less	
Closing stock	
Sales	
Transfers out	
	———
Total B	
	———
Amount used A–B	
	———

There may not be all of the above stages in any one case but it is important that they are not overlooked.

For both the quantity produced and quantity used calculation, the values may be physical or monetary or preferably both. The calculation is the same for both physical and monetary values. The heading No./wt./vol./£ merely indicates the options, which of course must relate to the issue under consideration.

Allocation to the appropriate enterprise

Allocation of quantity and cost are preferably entered directly on to a gross margin record in the relevant cost section. When a number of enterprises use raw material from a common source it may become necessary to use an allocation record which will enable regular checks on accuracy to be made against the total amount used record for that resource.

Fertiliser allocation 'N' to crops

	Total		Wheat		Barley		OSR	
Date	Tonne	£	Tonne	£	Tonne	£	Tonne	£
Total								

Entries to such a record should be made as soon after application/use as possible. Detail may be taken from a job sheet, diary or simply verbally from the responsible person. It is essential to check accuracy of allocation against stock records regularly.

The record may be developed to include field reference and hectares, but this will depend on individual opinion and requirements.

Internal transfer between enterprises

The records explained previously can be adapted to include detail of transfers in most instances, but it may be necessary to set up a custom made version for some circumstances. The important thing is to keep it simple and to ensure that transfers out of one enterprise are always transferred in to the appropriate enterprise (on paper).

Summary

There is no single 'right way' to record financial and physical data with a view to producing management information. It is essential to understand and plan in advance the requirements, to use basic principles and initiative in designing effective records and, finally, to ensure a disciplined approach to maintaining the records.

CHAPTER 11

Gross margins for Green Farm

In this chapter the accounts for Green Farm are broken down or analysed to produce gross margins for each enterprise and a gross margin account is drawn up which agrees with the profit and loss account which was considered in Chapter 9.

Fully detailed physical and financial records for the whole year would take up considerable space and would probably not be read and checked. Sample records are thus included to demonstrate method.

The gross margin data sheets shown are only one example of a suitable method of collecting input and output data. They are best maintained on a regular basis throughout the year. This will bring any recording errors or omissions to notice at an early stage, when they can most easily be rectified.

A further look at the profit and loss account for the year will serve as a reminder of content. It is helpful to mark and make notes on the account in the final stages when reconciliation with the gross margin account is carried out, so a spare copy should be available for this purpose.

The gross margin data sheets and accompanying gross margin calculations are supported by the sample physical records. It is essential that the reader check every stage of recording shown against the gross margin data and gross margin calculations in order to ensure full understanding of the process. For practical purposes it may be preferable to produce data sheets to suit individual requirements. A basic layout on A4 paper can be reproduced and stored in a binder. It is important to understand and stick to basic principles and to feel 'in command' of the situation.

The gross margin account must agree with the profit and loss account – failure so to do would indicate error or omission at some stage. There are three main areas to be checked for accuracy, these being **output calculation**, **cost calculation** and **internal transfer.**

In order to reconcile the two accounts, it is advisable to start with the profit and loss and determine that each component of, for example, output has been taken into consideration in the gross margin calculations. This is best done by neatly ticking each relevant entry

Green Farm Profit and Loss Account for year ending 31.3.X9

Valuation at 1.4.X8	£	£	Valuation at 31.3.X9	£	£
Dairy cows	45,000		Dairy cows	39,000	
Young stock	13,300		Young stock	13,700	
Barley (homegrown)	475		Barley (homegrown)	105	
Concentrate feed	300		Concentrate feed	1,650	
Seeds	3,210		Seeds	3,226	
Fertiliser	4,753		Fertiliser	5,540	
Fuel	600		Fuel	550	
Crop protection	1,764	69,402	Crop protection	1,170	64,941
Expenditure			*Revenue*		
Dairy cow feed	21,603		Milk	85,060	
Young stock feed	1,693		Calves	7,415	
Vet. and med.	2,828		Cull cows	10,540	
Livestock sundries	3,692		Barren heifers	1,486	
Seed	3,226		Winter Wheat		
Fertiliser	8,674		(Harvest X8)	28,313	
Crop protection	4,702		Winter Barley		
Fuel/oil/electricity	4,436		(Harvest X8)	6,930	
Machinery costs	8,716		Winter Oats		
Contract	730		(Harvest X8)	943	
Wages	20,948		Winter Barley		
General repairs	2,742		(Harvest X7)	475	
Rent/rates/water	14,590		Miscellaneous	1,602	142,764
General insurance/ Sundry fixed costs	3,542				
Administration/telephone	3,750		*Non-cash receipts*		
Bank interest and charges	5,643	111,515	Heifer – to house	450	
			Milk – to house	263	713
Depreciation					
Machinery	9,552				
Fixed equipment	152	9,704			
		190,621			
Profit		17,797			
		208,418			208,418

in the profit and loss account as it is checked against the gross margin calculation*. For example, for this particular case study, opening and closing valuations of dairy cows, purchase of replacement cows, sales of milk, cows and calves should all be checked to ensure that gross margin dairy output data agrees with profit and loss entries. By so checking and neatly marking every part of the profit and loss account which has been used in drawing up the gross margin account, unmarked profit and loss items are brought to attention for consideration. It may be that unmarked items have been overlooked in the

*In order to gain understanding, it is recommended that you copy the profit and loss account for this example and check and mark the entries as indicated. If the marking is done in the book itself it should be done lightly in pencil so that the marks can be erased.

Green Farm Gross Margin Account for year ending 31.3.X9

Fixed costs	£	£	*Gross margins*	£	£
			Livestock		
Fuel/oil/electricity	4,486		Dairy cows	62,846	
Machinery costs	8,716		Young stock	5,383	
Contract	730		Total	68,229	
Wages	20,948		Less forage costs	6,798	
General repairs	2,742		Total livestock	61,431	61,431
Rent/rates/water	14,590				
General insurance/sundry	3,542		*Cash crops*		
Administration/telephone	3,750		Wheat	21,273	
Herbicide	283	59,787	Barley	5,261	
Interest and bank charges		5,643	Oats	3,364	
Depreciation		9,704	Total crops	29,898	29,898
Total fixed costs		75,134	Miscellaneous		1,602
Profit		17,797			
		92,931			92,931

gross margin calculations, or that they do not form part of the main group of enterprises, as in the case of 'miscellaneous income' for this set of accounts.

Note that all valuations which appear in the profit and loss account are either taken to the gross margin output or cost calculations, or they are taken to calculation of fixed costs – see fuel in this case. All valuations must be taken forward to an output or cost calculation and so in the checking procedure every item must be marked – and therefore not appear on the gross margin account.

Internal transfers do not appear in the profit and loss account but they must be checked. Value 'given' by one enterprise must be balanced by value 'received' by another enterprise. Failure to balance transfers results in inaccurate gross margins and the gross margin account will not agree with the profit and loss account.

Reconciliation of cost is usually the area that requires most attention and so this is given in detail as follows. Check and mark each item to ensure that the procedure is fully understood.

Reconciliation of variable costs with the profit and loss account

Crops calculation – crops and fuel

	Seed	Fert.	Crop prot. sundries	(Fixed cost) Fuel + oil
	£	£	£	£
Opening stock	3,210	4,753	1,764	600
Purchases	3,226	8,674	4,702	4,436
	6,436	13,427	6,466	5,036
Less closing stock	3,226	5,540	1,170	550
Cost	3,210	7,887	5,296	4,486*

* Taken to 'Fixed costs' in gross margin account

Variable cost allocation

	Seed £	Fert. £	Crop prot. £	Crop sundries £
Wheat	1,768	2,081	3,050	141
Barley	632	367	705	70
Oats	56	92	305	–
Forage	754	5,347	322	375
Total cost	3,210	7,887	4,382	586

Crop sundries	586
** Fixed cost herbicide	283
	5,296

** Field records showed use of herbicides for control of couch which could not be reasonably allocated to the crop produced that year.

Cost calculation – purchased feed

	Conc. £
Opening stock	300
Purchases	21,603
	21,903
Less closing stock	1,650
Total cost	20,253

	Allocation £
Dairy cows	19,142
Dairy young stock	1,111
	20,253

Livestock Gross Margin Data Sheet

Enterprise: Dairy Young Stock Account Year Ending: 31.3.X9

Variable costs

Date	Quant./Des.	£	Date	Quant./Des.	£
Purchases Concs.			*Veterinary & Medicine*		
Opening stock		100.00	Apr.		00.00
Apr.		250.00	May		00.00
May		00.00	June		29.04
June		00.00	July		52.10
July		00.00	Aug.		00.00
Aug.		00.00	Sept.		74.00
Sept.		00.00	Oct.		32.08
Oct.		00.00	Nov.		116.80
Nov.		200.00	Dec.		179.00
Dec.		141.00	Jan.		71.13
Jan.		290.00	Feb.		89.00
Feb.		170.00	Mar.		74.85
Mar.		160.00			
Total available		1,311.00	Total		718.00
Less closing stock		200.00			
Cost		1,111.00			

Date	Quant./Des.	£	Date	Quant./Des.	£
Homegrown grain transferred in			*Livestock sundry*		
Apr.		55.00	May		43.00
May		55.00	Aug.		31.00
Sept.		55.00	Total		74.00
Oct.		110.00			
Nov.		110.00			
Dec.		110.00	*Milk for calves*		
Jan.		110.00	2,000 l @19.5p		390.00
Feb.		55.00			
Mar.		50.00			
Total		710.00			

Gross Margin – Livestock

Enterprise: Dairy Young Stock	Account Year Ending: 31.3.X9	
Enterprise output	£	£
Sales	1,486	
Transferred out	9,200	
Closing stock at valuation	13,700	24,386
Less		
Purchases – heifers	–	
– calves	200	
Transferred in – calves	2,500	
Opening stock at valuation	13,300	16,000
		8,386
Less variable costs		
Purchased feed	1,111	
Home produced – grain	710	
– milk for calves	390	
Vet. and med.	718	
Sundry	74	3,003
Gross margin before ded. of forage costs		5,383
Less forage costs		1,402
Gross margin after forage		3,981
Gross margin per ha (10.42)		382

Note: Output data was gained directly from the cash analysis or livestock numbers record, in order to keep recording tasks to a minimum. If a special data sheet were required it could be based on the example following.

Gross Margin Data Sheet — *Output*

Livestock Enterprise: .... Accounting Year ending: 31.3.X9

Output Data

Valuation Detail

Start of Year			*End of Year*		
No	@£ ...... hd	Total £.......	No	@£ ...... hd	Total £.......
Date	Quant./Des.	£	Date	Quant./Des.	£
Stock sales			Stock purch		
Total			Total		
Stock trans. out			Stock trans. in		
Total			Total		
Product sales			Sundry		
Total			Total		

Livestock Gross Margin Data Sheet

Enterprise: Dairy Cows Accounting Year Ending: 31.3.X9

Date	Quant./Des.	£	Date	Quant./Des.	£
Purchased concs.			*Vet. & med.*		
Opening stock		200.00	Apr.		152.81
Apr.		1,000.97	May		113.96
May		843.26	June		120.00
June		607.33	July		143.20
July		477.12	Aug.		146.99
Aug.		86.39	Sept.		260.57
Sept.		1,190.61	Oct.		180.00
Oct.		1,970.63	Nov.		147.27
Nov.		3,341.86	Dec.		292.12
Dec.		1,479.76	Jan.		172.18
Jan.		3,364.40	Feb.		185.75
Feb.		4,339.67	Mar.		195.15
Mar.		1,690.00	Total		2,110.00
Total available		20,592.00			
Less closing stock		1,450.00			
Cost		19,142.00			

Date	Quant./Des.	£	Date	Quant./Des.	£
Purchased bulk feed			*Livestock sundry*		
Apr.			Apr.		279.00
May			May		329.56
June			June		118.00
July			July		00.00
Aug.			Aug.		393.00
Sept.			Sept.		218.00
Oct.		148.00	Oct.		563.00
Nov.			Nov.		640.44
Dec.		152.00	Dec.		516.00
Jan.			Jan.		237.00
Feb.		398.00	Feb.		204.00
Mar.		120.00	Mar.		120.00
Total		818.00	Total		3,618.00
Homegrown grain transferred in					
Apr.		00.00			
May		220.00			
June		250.00			
July		350.00			
Aug.		440.00			
Sept.		150.00			
Oct.		110.00			
Nov.		220.00			
Dec.		220.00			
Jan.		110.00			
Feb.		65.00			
Mar.		74.00			
Total		2,209.00			

Gross Margin – Livestock

Enterprise: Dairy Cows — Accounting Year Ending: 31.3.X9

Enterprise output

	£	£
Sales – milk	85,060	
trans. to D.Y.S. & house	653	
Sales – cows	10,540	
calves	7,415	
Trans. in – heifers from D.Y.S.	2,500	
Closing stock at valuation 78 @£500	39,000	145,168
Less		
Purchased – cows	675	
Trans. in – heifers from D.Y.S.	8,750	
Opening stock at valuation 90 @£500	45,000	54,425
Enterprise output		90,743

Gross Margin – Livestock (continued)

Less variable costs		
Purchased – concs.	19,142	
bulk feed	818	
Trans. in – oats	2,209	
Vet. & med.	2,110	
Livestock sundry	3,618	27,897
Total gross margin before forage		62,846
Less forage costs		5,396
Total gross margin after forage		57,450
Gross margin per cow (av. no 83.5)		688
Gross margin per ha (forage area 40.08)		1,433

Gross Margin Data Sheet

Crop: Winter Oats   Ha: 8.0   Harvest Year Ending: 31.3.X9

Variable Costs

Date	Quant./Des.	£	Date	Quant./Des.	£
Seed			Fertiliser		
Oct.		56			
Total		56			92
Crop protection			Crop sundry		/
Total		350			–
Contract charges		/	Casual labour		/
Total		–			–

Output

Sales		Transfer for feed/seed	
Jan.	530	(from stockman's diary)	
Mar.	413	Total homegrown oats	2,919
Total	943	Total	2,919

Crops in store at year end – not sold – not transferred.	/
Total	–

Gross Margin

Crop: Winter Oats Ha: 8.0 Harvest Year Ending: 31.3.X9

Enterprise output	£	£
Sales 10 t		943
Transferred 31 t		2,919
Closing stock at valuation		–
Output		3,862
Less variable costs		
Seed	56	
Fertiliser	92	
Crop protection	350	
Crop sundry	–	
Contract charges	–	
Casual labour	–	498
Enterprise gross margin		3,364
Gross margin per ha	£420.50	
Yield per ha	5.12t	

Gross Margin Data Sheet

Crop: Winter Wheat Ha: 40.1 Harvest Year Ending: 31.3.X9

Date	Quant./Des.	£	Date	Quant./Des.	£
Seed			Fertiliser		
Total		1,768			2,081
Crop protection			Crop sundry		
Total		3,050			141
Contract charges		/	Casual labour		/
Total		–			–

Output

Sales		Transfer for feed/seed	
	4,280		
	8,946		
	6,640		
	5,707		/
	2,740		
Total 283 t	28,313	Total	–

Crops in store at year end – not sold – not transferred	/
Total	–

(continued)

Gross Margin

Crop: Winter Wheat	Ha: 40.1	Harvest Year Ending: 31.3.X9

Enterprise output	£	£
Sales 283 t		28,313
Transferred		–
Closing stock at valuation		–
Output		28,313
Less variable costs		
Seed	1,768	
Fertiliser	2,081	
Crop protection	3,050	
Crop sundry	141	
Contract charges	–	
Casual labour	–	7,040
Enterprise gross margin		21,273
Gross margin per ha	£530.50	
Yield per ha	7.05t	

Gross Margin Data Sheet

Crop: Winter Barley	Ha: 11.6	Harvest Year Ending: 31.3.X9

Date	Quant./Des.	£	Date	Quant./Des.	£
Seed			Fertiliser		
Total		632	Total		367
Crop protection			Crop sundry		
Total		705	Total		70
Contract charges		/	Casual labour		/
Total		–			

Output

Sales		Transfer for feed/seed	
Oct.	2,480		
Nov.	1,228		/
Mar.	3,222		
Total 73t	6,930	Total	–

Crops in store at year end – not sold – not transferred	
Total 1 tonne	105

(continued)

Gross Margin

Crop: Winter Barley	Ha: 11.6	Harvest year ending: 31.3.X9
Enterprise output	£	£
Sales 72t		6,930
Transferred		–
Closing stock at valuation 1t		105
Output		7,035
Less variable costs		
Seed	632	
Fertiliser	367	
Crop protection	705	
Crop sundry	70	
Contract charges	–	
Casual labour	–	1,774
Enterprise gross margin		5,261
Gross margin per ha	£453.53	
Yield per ha	6.29t	

Forage Variable Costs

Forage Area: 50.5 ha — Accounting Year Ending: 31.3.X9

Date	Quant./Des.	£	Date	Quant./Des.	£
Seed			Casual labour		/
		754			–
Fertiliser			Sundry receipts		/
		5,347			–
Crop protection			*Cost summary*		
			Seed		754
			Fertiliser		5,347
		322	Crop protect.		322
			Contract charge		–
			Sundry		375
Contract charges		/	Casual labour		–
		–	Total variable costs		6,798
			Less sundry receipts		–
Sundry			Less increase in val. OR Plus decrease in val.		–
		375	Total forage costs		6,798
			Per ha		134.6

ALLOCATION OF FORAGE COSTS

It is very difficult to allocate forage costs accurately when different classes of stock graze the same area – who can possibly tell the benefit each animal gains from grazing? The burden of recording precise stock numbers relative to grazed area would outweigh the gain – and anyway it simply would not get done in most cases.

An acceptable compromise is to calculate average stock numbers and convert to livestock units in order to calculate total forage area for the different groups. Figures for Green Farm would look like this:

Livestock units	Av. No.		Value	LSUs
Cows	83.46	×	1	83.5
Dairy Y.S. over 1 yr	21.15	×	0.6	12.7
Dairy Y.S. 0–1 yr	22.46	×	0.4	9.0
Total livestock units				105.2
Allocation of cost				
Dairy cows	83.5 / 105.2	×	£6,798	£5,396
Dairy young stock	21.7 / 105.2	×	£6,798	£1,402
				£6,798
Forage area				
Dairy cows	83.5 / 105.2	×	50.5 ha	40.08 ha
Dairy young stock	21.7 / 105.2	×	50.5 ha	10.42 ha
				50.50 ha

LIVESTOCK NUMBERS

Monthly head count Year ending: 31.3.X9

Month	Cows	DYS over 1 yr	Dairy calves 0–1 yr	Bull calves 0–1 yr	Beef
Opening no.	90	19	19		
April	89	19	13		
May	90	21	13		
June	95	21	14		
July	95	29	26		
Aug.	80	28	28		
Sept.	80	24	24		
Oct.	76	23	25		
Nov.	78	21	26		
Dec.	77	19	26		
Jan.	78	17	26		
Feb.	79	17	26		
Mar.	78	17	26		
Total Nos.	1,085	275	292		
Av. Nos.	83.46	21.15	22.46		

Both a monthly head count and a monthly livestock movement record are maintained for the farm in this example.

Numbers are recorded on the last day of each month. Minute breakdown into strict age groups has been avoided due to the practical difficulty of counting and the necessity for progressive transfer.

Extracts from a 'sample' stockman's diary are shown below and the detail is then shown in a livestock movement record for May.

Stockman's Diary (extract) May 19X8

6th May 1 heifer calved – heifer calf
8th May Cow no. 47 calved – heifer calf
9th May Cow no. 63 calved – bull calf
12th May Cow no. 98 calved – heifer calf
16th May 3 one-year-old heifer calves moved to Dairy Young Stock unit
17th May 1 bull calf sold (1–2 weeks old)

Livestock Movement Record May 19X8

	Cows in milk	Cows dry	D.Y.S. over 1 yr	D.Y.S. 0–1 yr	B. calves 0–1 yr
Opening stock nos. PLUS	83	6	19	13	
Purchased					
Bred				1/1/1	1
Trans. in	1 heifer 1 cow(47) 1 cow(63) 1 cow(98)		3		
Total A	87	6	22	16	1
LESS Sold					1
Died					
Trans. out		1/1/1	1	3	
Total deductions B	–	3	1	3	1
Closing stock A–B	87	3	21	13	–

This chapter presents a rather daunting amount of information to the newcomer. If at this stage a full grasp of the procedure has not been gained it may be helpful to the reader to revise, taking one enterprise at a time, and gradually build the complete picture – and confidence.

A 'bird's eye' view of the procedure is important to full understanding, regardless of the type and scale of business. Consider the basic principles relating to internal transfer, cost calculation and output calculation. Visualising these principles in the context simply of 'goods' rather than the complexities of a specific business can help to gain this overall view.

CHAPTER 12

Planning for the future: theory

Everyone makes plans for all aspects of life and business is no exception. The way in which plans are made is quite another matter. Some are carefully prepared and reasonably accurate whilst others are simply muddled optimism, but then we are reminded that 'the best laid plans of mice and men . . .'.

Why plan business activities when so many things can go wrong? There will be interest rate changes, tax changes, rent, rate, wage and raw material cost increases to mention just a few. Why not just muddle on and hope for the best?

Vague planning based only on hope is a product of muddled thinking. Lack of clear aim or target is a product of uncertainty and leads to more uncertainty. Perhaps the greatest disadvantage of a lack of planning is that serious problems are only discovered when it is too late to do anything. It is a bit pointless bolting the door after the horse has gone. A plan allows progress to be monitored.

On what basis do we plan then? In business as in other spheres of life, planning is most reliably based on what is known from past experience.

HISTORICAL ACCOUNTS – A BASIS FOR PLANNING

Clearly, what is known of past business activity will be permanently recorded in the accounts. It is on these accounts that plans for future activity must be based – published information of the achievements of others is only of limited value, except maybe to act as a spur to higher levels of achievement. Past performance of the business must be the basis for planning because of the major influences of people, location and source of funding. Known qualities of these influences must be taken into account when the question is asked, 'What have we achieved and can we do better?'

Historical accounts form a basis, but which historical accounts? Are the most recent accounts the most accurate reflection of past achievements or would earlier accounts give a more reliable picture? The truth is that it is unlikely that any one set of accounts provides a sufficiently representative record of business achievement. Vagaries of weather, fashion, political upsets etc. can all lead to fluctuations for good or ill in the short term and so a more balanced basis for planning must be used.

Normalised accounts

There is no perfect basis for business planning because, fortunately, we do not know what the future holds in store. A reasonable solution and one which is widely used is the **normalised** account. A normalised account consists of an average of physical inputs and outputs – for example, fertiliser application and yield of grain, or concentrate feed and yield of milk – taken over two or three years to which current prices and values are attached. The account should also be 'normal' in terms of overall plan, or scale and combination of individual enterprises. An account is therefore produced from which abnormal fluctuations are removed and which reflects current values.

The gross margin account should form the basis of the normalised account – the profit and loss account is too vague. Individual margins are normalised as described above and fixed costs are also adjusted as necessary, to represent a normal year (if there could be such a thing!). The end result is a gross margin account which purports to represent the result of trading in a 'normal' or average year and provides a sound basis for planning and budgeting.

PLAN AND BUDGET

Planning concerns the physical, budgeting attaches finance to the plan. You plan to go on holiday, where, when, how; and the holiday budget indicates the likely cost of your plan. Likewise with business plans and budgets.

Complete budgets

A complete budget, as the name implies, budgets for the whole business and is used to predict likely capital requirements and profit. Financial control of a business causes more problems than any other single issue; a cash flow budget based on the complete budget predicts likely capital requirement on a periodic basis throughout the year. More on cash flow later.

As mentioned earlier, a plan must precede a complete budget. A physical plan is produced bearing in mind the constraints relative to land, labour and capital. For agriculture and horticulture the problem of disease control through the crop rotation must be considered. For intensive farming, horticulture or manufacturing, productive capacity must be taken into consideration in the plan. The other major physical input is labour – a costly resource which must be used effectively. A seemingly attractive plan based on high gross margin enterprises which have coinciding high peak labour requirements could lead to difficulties and possibly failure.

When the plan has been produced, including all detail of physical inputs and outputs, then prices can be applied to this data. Care should be taken to determine likely price increases for inputs but it is wise to take a rather pessimistic view on value of outputs – far better to underestimate than to be too optimistic and face disappointment and a deficit at the end of the year.

Partial budgets

Partial budgets are used to predict the likely effect on profit when considering change or substitution of an enterprise. Gains and losses are balanced to indicate likely results. The budgets may be based on gross margins already calculated or they may be fully detailed. Additional capital investment in buildings or equipment, for example, is included in the budget as an annual depreciation charge.

Basic principles of partial budgeting

Four potential effects must be taken into consideration in a partial budget: revenue lost and additional cost, revenue gained and cost saved.

Revenue lost as a result of giving up an enterprise is effectively a cost to the business.

Additional cost includes the additional fixed costs when the budget is based on gross margins. When a fully detailed budget is drawn up all of the new or replacement enterprise costs are included.

Revenue gained from the new enterprise is on the 'plus' side of the budget and may be shown as a gross margin or full detail of output, dependent on style of budget.

Cost saved by giving up an enterprise is also a 'plus' factor in the budget. The cost saved would be real reduction of fixed costs in a gross margin based budget or of enterprise costs and fixed costs in a fully detailed budget. It is important that a realistic assessment of saving in fixed costs is made; machinery or buildings made redundant may not result in real saving and only a genuine saving in labour costs, such as reducing the regular full or part-time work force, should be considered.

See example on page 114.

NARRATIVE

It is essential that with each partial budget there is a brief but clearly detailed description of the proposal.

e.g. Budget to replace 5 ha of winter barley with 5 ha of maincrop potatoes. Contractors to plant. Local labour available for harvesting.

OR Budget to increase the breeding flock of sheep by 100 ewes and replace 10 ha of cereals with grass. No additional labour required. Present winter housing adequate for increased numbers but extra feed troughs and hayracks required.

Partial budget – based on gross margins

Revenue lost	£	*Revenue gained*	£
(For gross margin of enterprise given up)		(For gross margin of new or increased size of enterprise)	
Additional cost		*Cost saved*	
(Additional fixed costs)		(Real saving in fixed costs)	
Balance being Extra income		OR Reduction in income	

Partial budget – fully detailed

Revenue lost	£	*Revenue gained*	£
Detail of output for enterprise given up		Detail of output for new or increased size of enterprise	
Additional cost		*Cost saved*	
Variable costs for new or increased size of enterprise plus additional fixed costs		Variable costs for enterprise given up plus real saving in fixed costs	
Balance being Extra income		OR Reduction in income	

THE CASH FLOW BUDGET

Financial control of business is the area where the majority of problems are encountered and lack of control leads to many business failures.

Bank managers must be assured that finance is under control, creditors must be paid at the right time and debtors must be pressed to settle within acceptable time limits. The business plan must be taken through to completion without premature disposal of produce and certainly without disposal of fixed assets such as breeding stock. This all presents a difficult balancing act, particularly when there is heavy reliance on borrowed capital. Only the foolhardy will venture through such a jungle without a route map – the cash flow budget.

What exactly is cash flow?

Cash flow is the difference or balance between inflow (receipt of cash) and outflow (payment of cash) over a stated period. If £1,000 is received and £1,200 paid out over a period, there is a negative flow of £200. Cash flow figures for periods within the financial year are accumulated with the opening bank balance to indicate bank balance for the period. There is little benefit in calculating actual cash flow if no cash flow budget has been produced.

A simple cash flow – for 3 months, quarters or other periods of time

(Brackets indicate negative figures)

	No.	Month/qtr 1	No.	2	No.	3
Receipts (detail)						
Total receipts A		1,000		1,500		1,200
Payments (detail)						
Total payments B		(1,200)		(1,100)		(1,600)
Cash flow A–B		(200)		400		(400)
Opening bal. 100 Accumulated bal.		(100)		300		(100)

Money in the bank at the start of the cash flow was £100. In the first period payment exceeded receipt by £200, indicated in brackets since it is a negative amount. The effect of the (£200) on bank balance gives (£100) when the opening balance is taken into consideration.

In the second period receipt exceeds payment by £400 and this when accumulated with the bank at the previous period results in a new bank balance of £300, i.e. (£100) + £400 gives £300.

In the final period payment exceeds receipt by (£400). Bank balance of £300 in the previous period accumulates to give a final balance of (£100), i.e. £300 + (£400) gives (£100).

Plus and minus signs are often used rather than brackets for negative figures. Brackets have the advantage in that they are more easily seen and there is no need to add a sign to positive figures.

Cash flow prediction

At best, the prediction is nothing more than just that, or in other words an informed guess. The chief benefits are found in that likely danger areas are brought to notice in advance, e.g. requirement for increased borrowing at a particular point in time, and targets or 'marker posts' are created against which actual cash flow can be monitored. The targets help to keep the business to plan for the year, deviation becomes obvious and so corrective action can be taken before it is too late.

CONSTRUCTING THE CASH FLOW BUDGET

A plan must, as always, precede the budget. Plans are made to buy and sell goods throughout the year according to need and production cycles. The difficult part is coping with the facts that time of payment usually differs from the time of transaction, some debtors are not too predictable and debtors and creditors from the previous year must be included in the budget, to mention just a few problems.

The complete budget forms the basis for cash flow prediction. Individual total purchases and sales from the budget must be broken down over the year to fit the planned pattern of transactions and production. The most satisfactory way to do this is to take one item at a time and plan the times of purchase and the times of payment over the total period – this may be done in a way which suits the person concerned but it must be orderly and detailed. Detail of predicted payment and receipt is then entered on the cash flow form in the appropriate place. When all items have been entered horizontally across the cash flow form, vertical columns are totalled and cash flow is calculated.

Financial detail must be supported by physical information where appropriate. Quantity or unit price of goods to be bought or sold can be compared to what actually takes place. If the value of (e.g.) wheat sold is lower than the predicted price, it is important that the reason for the difference can be determined. If there is no information on planned quantity and unit price the exercise becomes vague and unsatisfactory.

INTEREST CHARGES

Interest is calculated on negative bank balance according to the length of period to which the cash flow relates. If cash flow is calculated monthly the amount charged on a negative balance is 1/12 of the annual rate, e.g. 1/12 of 18% = $1\frac{1}{2}$%. The most common way of entering interest charges is to calculate the amount due on a negative bank balance – or overdraft – and enter it as a payment in the following period. It may be considered desirable to accumulate interest charges and enter them in the cash flow at the point when charges would normally be made by the bank.

	Month 1 £	Month 2 £	Month 3 £
Receipts			
(Detail)			
Total receipts A	1,000	1,500	2,000
Payments			
(Detail)			
Interest charge		15	24
Total payments B	2,000	2,100	2,800
Cash flow A–B	(1,000)	(600)	(800)
Cumulative (op. bank o)	(1,000)	(1,600)	(2,400)
	Interest 18% p.a. £1,000 × 1½% = £15	£1,600 × 1½% = £24	£2,400 × 1½% = £36 (to be entered in the next period)

DETAILED CASH FLOW

Cash flow for period .... to ....

	Period 1 Month £	Period 2 Month £	Period 3 Month £
Trading receipts Total A	5,000	8,000	6,000
Trading payments Total B	6,000	7,000	7,000
Trading net cash flow A–B	(1,000)	1,000	(1,000)
Interest charges	–	(54)	(49)
Trading N.C.F. and int. charges C	(1,000)	946	(1,049)
Capital receipts	–	–	–
Capital payments	(2,000)		–
Capital cash flow D	(2,000)	–	–
Private receipts	–	–	–
Private drawings (inc. tax)	(600)	(600)	(600)
Private cash flow E	(600)	(600)	(600)
Cash flow C + D + E	(3,600)	346	(1,649)
Cumulative C.F. (op. bal. o)	(3,600)	(3,254)	(4,903)

Monthly cash flow. Interest 18% = 1½% per month

Details of payments and receipts are omitted in order to concentrate on basic principles.

The examples given so far set out the basic principles of calculating cash flow. Receipts and payments should be divided into broad categories as well as into specific detail. Trading receipts and purchases should be separated from capital and private items.

Trading receipts and payments represent the cash result of trading activities and the resulting cash flow is generally known as the **trading net cash flow.**

Capital receipts and payments are made in a supportive role to the business for provision of machinery, buildings and equipment, as will be fully understood by now. Influences and decisions regarding these receipts and payments are different, even though connected, from trading items and so the two should not be confused.

Private receipts and payments are not business items but rather introduction of funds from private sources (private receipts) or drawings for private use (private payments). There is a clear need to separate off these items since they may have considerable influence on cash flow irrespective of business activities. Drawings greater than is desirable for the business could put the whole operation at risk and so it is important that detail is shown separately.

TOTAL CASH FLOW

Total cash flow or, when it concerns a year, annual cash flow, should be calculated primarily as a means of checking the accuracy of individual cash flow periods. Cash flow for the final period must be the same as the total (or annual) cash flow. If there is a discrepancy it must be found by careful checking – an apparently small discrepancy sometimes hides a greater error.

Total for cash flow calculated on page 117 – this would normally be shown in either the first column or the last column of the cash flow matrix.

	£
Total receipts	19,000
Total payments	20,000
Trading N.C.F.	(1,000)
Interest	(103)
Capital	(2,000)
Private	(1,800)
Cash flow	(4,903)
Cum. C.F. (inc. o.b.)	(4,903)

Reconciling cash flow with total cash flow

As always, it is essential to be methodical when checking figures. In this instance there are two main areas to check, horizontal breakdown of amounts between cash flow periods and vertical arithmetic. The following four steps should enable any problem to be located:

1) Check arithmetic of total cash flow.
2) Check horizontal breakdown for every item, i.e. do amounts add up to agree with the total?

3) Check arithmetic vertically for every cash flow period.
4) Check arithmetic for accumulation of cash flow.

CASH FLOW AND PROFIT

Cash flow does not indicate profit for the period, it merely shows inflow and outflow of cash. The influence of opening and closing debtors and creditors, valuation change, depreciation and notional receipts must all be taken into consideration for calculation of profit. Consider the effect of each stage in the procedure set out below, drawing on knowledge gained earlier.

	£
Trading net cash flow for period	
Annual total	
Notional receipts (produce consumed, house rent, private use of car etc)	
Closing valuation • Livestock	
• Harvested crops	
• Tillages and growing crops	
• Stores	
Closing debtors	
Opening creditors	
SUB TOTAL (a)	———
Opening valuation • Livestock	
• Harvested crops	
• Tillages and growing crops	
• Stores	
Closing creditors	
Opening debtors	
SUB TOTAL (b)	———
Profit (a−b) before depreciation	———
Depreciation on buildings and fixtures	
Machinery depreciation	
Trading profit after depreciation	═══

VAT in the cash flow

All VAT paid out by a VAT registered business is reclaimable (with the exception of exempt and partially exempt businesses). Prompt recording and return of the VAT 100 ensures that VAT has a minimal effect on cash flow. Cash recording of VAT allows collection of tax on sales (when applicable) before payment to Customs and Excise, which may give a brief advantage to cash flow. All things considered, it is generally easier to ignore VAT in the cash flow and concentrate on the real transactions and movement of cash. If this rule is followed, all payments and receipts will be entered excluding VAT and VAT reclaimed or paid to Customs and Excise will not be entered.

If VAT entry to cash flow is required, then it may most suitably be done by including a 'VAT received from sales and C & E' entry in 'Trading receipts' and 'VAT paid to suppliers and C & E' in 'Trading payments'. All payment and receipt for goods and services are shown excluding VAT; this is helpful when drawing up a budget cash flow and when comparing budget with actual cash flow.

PUTTING THE PLAN INTO OPERATION

It is perhaps something of a miracle if all goes according to plan. Transactions are carried out and work is done with the intention of sticking to the guidelines, but no sooner does the action commence than the scene starts to change. Prices increase or decrease, technical and labour problems arise, the market changes and seemingly in no time the business is off course. What is to be done?

Monitoring the plan and cash flow

Attitude of mind is all important. Whilst efforts should be made to keep to the best aspects of the plan, a flexible outlook which allows constant revision is necessary. Priorities must be determined; there is no point in wasting time on smaller items of receipt or payment or on those which cannot be adjusted, such as rent and rates.

Inputs and outputs over which management may exercise control are the key areas. Are these receipts and payments near to the plan? If not, why not? Take a specific example of output, e.g. milk sales. If receipt for the month is down, the amount of variance is determined and then analysed to discover why. It will be due to one or more of three reasons:

variance from plan of main productive factor – i.e. number of cows
variance from budgeted yield per cow
variance from budgeted price per litre.

Payment variance, e.g. variance in concentrate cost for the cows, will be due to one of the following three reasons:

variance from plan – less/more cows to feed
variance on budgeted amount used – less/more concentrates per cow
variance on budgeted price – less/more per tonne.

Some will consider that this detailed approach is a bit 'over the top'. The important point is that areas of major influence should be monitored carefully. It takes only a few minutes to make the analysis described above if the information is available. In many cases it will be said 'Well I know I've got more/less cows than I planned' or 'I know I paid a bit more for the cake so there's no need to go to all that trouble'. In such cases the answer may well be to hand, but probably lacking detail.

A major benefit of cash flow budgeting and monitoring is that it creates an awareness of what is happening. Decisions to adjust or change can be made in good time rather than as a panic measure.

REVISION OF THE CASH FLOW BUDGET

Change will inevitably occur and so the cash flow budget will need to be changed if it is to have any relevance after the first few periods.

Revision of manual cash flow calculation is a tedious and frustrating task. It is often at this stage that the whole thing gets a bit overwhelming and is abandoned. If the calculation is to be done manually, use a pencil rather than pen to allow erasure and stick to priorities in the budget.

The computer is perhaps at its most useful to small businesses for calculation (or computing) of cash flow. A cash flow matrix is very easily set up on a spreadsheet program and the donkey work of writing and arithmetic is taken away – all that is necessary is revision of the offending figures. Purchase of a computer may well be justified for this purpose alone.

CHAPTER 13

Planning for the future: practice

The manager of Green Farm, after receiving his annual accounts a little late, decides to draw up a budget and cash flow for the year commencing in October. By this time he knows what his grain crop has produced and, conveniently, the plan will cover a full harvest year.

The process of normalising historic accounts in this case is not shown. In fact, the manager knows the farm well and has constructed budget gross margins based on his knowledge of past performance. The normalising process, extracting information from historical records, must be followed more methodically if the person drawing up the budget has not had first hand involvement in running the farm.

The overall plan is little changed except for slight simplification of the cereals area. Bank overdraft has increased through the summer months and the owner has requested a cash flow budget so that he can review the situation.

Plan for the year

Cropping for harvest year 19XX		*Stocking*	
W. wheat	50.00 ha	Dairy cows	84
W. barley	10.00 ha	Dairy youngstock	
Grass	50.20 ha	1–2 yrs	21
	110.20	0–1 yr	23

Predicted output and costs can be seen in the enterprise gross margins.

GROSS MARGINS – BUDGET

The gross margin budget account is supported by enterprise gross margins. Sufficient detail is provided in the example to enable reconciliation of the gross margin budget with the cash flow budget, but otherwise detail is entered sparingly. Clearly in a real life situation plans and budgets should be fully detailed but this is for the purpose of comparison with what actually happens rather than the vain hope of providing a precisely accurate budget.

Sensitivity factors would be used to determine the effect of price changes of sales and purchases. The reader may calculate the effect of variation in the price of milk per litre, feed per tonne and calves or other stock at a per head or per kg price. Small realistic changes often have a surprising effect on the total inputs or outputs of an enterprise.

Green Farm Budget 1.10.X9 – 30.9.XX

Fixed costs	£	£	*Gross margins*	£	£
Fuel/oil/electricity	4,400		Dairy cows	62,091	
Machinery costs	7,000		Youngstock	6,749	68,840
Contract	800				
Wages	21,400		Wheat	27,340	
General repairs	3,200		Barley	4,310	31,650
Rent/rates/water	15,200				
General insurance/sundry	2,450				
Administration/telephone	3,250	57,700			
Interest	4,828				
Depreciation	10,993	15,821			
		73,521			
Profit		26,969			
		100,490			100,490

Depreciation of machinery and fixed equipment

No machinery was purchased or sold between April and September and so for simplicity the April valuation is taken as a basis for calculation of depreciation, since it is a notional expense appearing in a budget.

Machinery account (budget)	£
W.D.V. at 1.10.X9	38,206
Purchases:	
Tractor & power harrow	16,000
	54,206
Sales	–
	54,206
Less depreciation @20% p.a.	10,841
W.D.V. at 30.9.XX	43.365

Fixed equipment account (budget)	
W.D.V. 1.10.X9	760
Depreciate at 10% of cost £1,520	152
	608
Total depreciation	10,993

Budget Gross Margin
Dairy Cows — *Year Ending 30.9.XX*

	£	£
Enterprise output		
Sales – milk 4,657 hectalitres	87,250	
cows 22	9,750	
calves 59	7,690	
Transfer to Dairy Youngstock 24 calves @£100	2,400	
Closing stock at valuation	42,000	149,090
Less		
Transfer in 19 heifers @£700	13,300	
Opening stock at valuation	42,500	55,800
Enterprise output		93,290
Less variable costs		
Purchased concentrates 132t	19,180	
Homegrown grain 20t barley (transfer in)	2,000	
Vet. and med.	2,070	
Livestock sundry	2,510	
Variable costs		25,760
Total gross margin before forage costs		67,530
Less forage costs		5,439
Total gross margin		62,091
Gross margin per cow (average 84)		739
Gross margin per hectare (39.85 ha)		1,558

Budget Gross Margin *Dairy Youngstock*		*Year Ending 30.9.XX*
Enterprise output	£	£
Sales – barren heifers 2	1,100	
Transfer to dairy herd		
19 heifers	13,300	
Closing stock at valuation	16,000	30,400
Less		
Transfer in 24 calves @£100	2,400	
Opening stock at valuation	16,200	18,600
Enterprise output		11,800
Less variable costs	£	£
Purchased concentrates	1,870	
Homegrown grain 10t barley		
(transfer in)	1,000	
Vet. and med.	330	
Livestock sundry	440	
Variable costs		3,640
Total gross margin before forage costs		8,160
Less forage costs		1,411
Total gross margin		6,749
Gross margin per hectare (10.35 ha)		652

Budget Gross Margin *Winter Wheat 50 ha*		*Year Ending 30.9.XX*
	£	£
Enterprise output		
Sales		
Transfer for home use		
Closing valuation 350t	36,600	
Enterprise output		36,600
Less variable costs		
Seed	2,300	
Crop protection	4,150	
Fertiliser	2,810	
Variable costs		9,260
Total gross margin		27,340
Gross margin per hectare		547

Budget Gross Margin		*Year Ending 30.9. XX*
Winter Barley 10 ha	£	£
Enterprise output		
Sales		
Transfer for home use		
Closing valuation 60t	5,850	
Enterprise output		5,850
Less variable costs		
Seed	500	
Crop protection	600	
Fertiliser	440	
Variable costs		1,540
Total gross margin		4,310
Gross margin per hectare		431

Forage costs
Budgeted total forage cost is £6,850
Allocation of forage cost
Livestock units – from plan

	Av. no	Value	L.S.U.s
Cows	84	×1.0	84
D.Y.S. over 1 yr	21	×0.6	12.6
D.Y.S. 0–1 yr	23	×0.4	9.2
			105.8

Allocation of cost			
Dairy cows	84 / 105.8	×£6,850	£5,439
Dairy youngstock	21.8 / 105.8	×£6,850	£1,411
Forage area			ha
Dairy cows	84 / 105.8	×50.2	39.85
Dairy youngstock	21.8 / 105.8	×50.2	10.35

CHAPTER 14

The cash flow budget for Green Farm

The complete budget having been drawn up, we can now produce the cash flow budget. A quarterly cash flow budget has been chosen to demonstrate the principle since it is compact and covers the complete year. Underlying principles are identical whether cash flow is constructed on a daily, weekly, monthly or quarterly basis. The shorter the period, the greater the likelihood of discrepancy between budget and actual due to the difficulty in predicting time of payment.

Physical detail is confined to those major items which are influenced by day to day management such as concentrate feed for the cows, milk, cull and calf sales. Major purchases such as fertiliser will be planned and purchased in bulk in advance – the number of hectares is unlikely to vary during the year but cow numbers and performance are less predictable.

When first attempting to produce a real cash flow budget it is important to be realistic and pay attention to detail but not to get bogged down with it to such an extent that you give up. Sophistication can always be added later as 'fine tuning'.

When the budget cash flow was drawn up, debtors and creditors at the beginning and end of the year had to be taken into consideration. These were as follows:

		£
Debtors		
Milk 1st October 19X9	40,000 litres	6,800
30th September 19X0	42,000 litres	7,350
Creditors		
Dairy cow feed		
1st October 19X9	10 tonnes	1,500
30th September 19X0	5 tonnes	750
Valuations		
October 1st 19X9		
Dairy cows (85)		42,500
Dairy youngstock		16,200
Wheat in store 300 t – for sale		31,600
Barley in store 30 t – for sale		2,850
30 t – for feed		3,000

	£
October 1st 19X0	
Dairy cows (84)	42,000
Dairy youngstock	16,000
Wheat in store 350 t	36,600
Barley in store 60 t	5,850

Capital purchases during the budget year	
Tractor (used)	£10,000
Power harrow (new)	£ 6,000
Drawings and tax	£12,000
Bank interest	....

Bank interest is budgeted at 18 per cent p.a., calculated on the closing balance of the previous quarter. This lacks accuracy due to possible fluctuations within the three months, but provides a sufficient guide.

PLANNING THE BUDGET CASH FLOW

Receipts

Milk sales It is common knowledge to dairy farmers that payment, by credit transfer, occurs about the middle of the month following the month of delivery. This means that receipts for milk always relate to the previous month for cash flow purposes.

Receipts for the first quarter, being October, November and December will include milk delivered in September but not December delivery. Likewise for the following three quarters.

Planning for cash flow in this instance must start with an estimate of milk yield and then price per litre is predicted. Dairy farmers are fortunate in that a great deal of assistance is available with regard to yield prediction and seasonal milk prices – other industries are not so fortunate.

Yields to which **receipts** relate are predicted as follows. Total values are rounded slightly for convenience:

	litres	*pence/litre*	*total £*
Oct./Nov./Dec.	90,500	20	18,100
Jan./Feb./Mar.	148,158	19	28,150
Apr./May/Jun.	151,250	16	24,200
Jul./Aug./Sept.	73,864	22	16,250
	463,772		86,700

The remaining receipts are shown on the cash flow itself (opposite page). A series of notes and jottings would normally be made prior to actual entry on the cash flow.

Budget cash flow

TRADING	Oct.-Dec. X8 Budget		Jan.-Mar. X9 Budget		Apr.-Jun. X9 Budget		Jul.-Sept. X9 Budget		Quantity	Total
Receipts	Quantity	£	Quantity	£	Quantity	£	Quantity	£		
Milk	90,500	18,100	143,158	28,150	151,250	24,200	73,864	16,250	463,772	86,700
Calves	13	5,850					9	3,900	22	9,750
Cull cows	49	6,440					10	1,250	59	7,690
Barren heifers							2	1,100	2	1,100
Wheat (19X9 Crop)	200	20,600	100	11,000					300	31,600
Barley (19X9 Crop)	30	2,850							30	2,850
Sub Total (a)		53,840		39,150		24,200		22,500		139,690
Payments										
Dairy cow feed	40	5,980	55	8,600	24	3,500		1,850		19,930
Dairy young stock feed	6	770	7	900	2	200				1,870
Vet. & med.		600		600		600		600		2,400
Livestock sundries		750		850		650		700		2,950
Seed		2,800						600		3,400
Fertiliser		3,500		4,250		1,500				9,250
Crop protection		2,200		500		2,300				5,000
Fuel,oil,elec		1,100		1,100		1,100		1,100		4,400
Machinery costs		1,500		2,000		1,500		2,000		7,000
Contract	Hedging	400			F.Y.M.	400				800
Wages		5,200		5,400		5,000		5,800		21,400
General repairs		800		800		800		800		3,200
Rent,rates,water		3,800		3,800		3,800		3,800		15,200
Gen. ins., sundry fixed costs		400		400		1,250		400		2,450
Admin., telephone		800		800		800		850		3,250
Sub Total (b)		30,600		30,000		23,400		18,500		102,500
Net cash flow (a−b)		23,240		9,150		800		4,000		37,190
Interest on previous c.bal		1,485		1,068		943		1,332		4,828
Trading net cash flow	(A)	21,755		8,082		143		2,668		32,362
PERSONAL										
Receipts, Sub Total (c)										
Drawings		2,500		2,500		2,500		3,000		10,500
Tax										
Any other payments				2,800				1,500		4,300
Sub Total (d)		2,500		5,300		2,500		4,500		14,800
Personal cash flow	(B)	(2,500)		(5,300)		(2,500)		(4,500)		(14,800)
CAPITAL										
Grants										
Mach. sales										
Sub Total (e)										
Buildings etc.										
Machinery	Tractor	10,000			P. Harrow	6,000				16,000
Other capital repayments										
Sub Total (f)		10,000				6,000				16,000
Capital cash flow (e−f)	(C)	(10,000)				(6,000)				
Net cash flow (A+B+C)		9,255		2,782		(8,643)		(1,832)		1,562
Cumulative balance		(23,745)		(20,963)		(29,606)		(31,438)		(31,438)
Opening balance	(33,000)									

Payments

Purchased concentrates Purchased concentrate feed is an expensive item in any intensive livestock enterprise and one over which careful control must be exercised. In this case, planned payment must include creditors at the start whilst payment for December feed will not be made until January under normal terms of trade. Likewise for the remaining three quarters of the year, which means that payment for the final month's feed will not take place soon enough to be entered on the cash flow for the year.

Checking the cash flow

If you are fortunate, the net cash flow for the final period will agree with that calculated from total receipts and payments. Spend a little time cross checking the example given to ensure that you understand the principles.

RECONCILING CASH FLOW WITH BUDGET PROFIT

It is essential that these two predictions do agree – if they don't, troubles will commence prematurely!

Only the trading net cash flow is taken into consideration for this purpose, since capital payments and receipts go to the depreciation account and personal or private transactions go to drawings. Adjustment to cash flow is made in two stages; stage one adds to cash flow, stage two deducts from cash flow.

During construction of the budget cash flow opening debtors and creditors are included but closing debtors and creditors are not. In the first stage, closing debtors are added because they belong to sales for the year, opening creditors are added back because they have been entered as a payment for the year but do not 'belong' to that year. In the second stage, closing creditors are included since they 'belong' to the budget year but have not been included in cash flow, and opening debtors are entered in this deduction stage because they have been included in cash flow (thereby improving it), but do not belong.

Closing valuations are included, or added to cash flow in the same way that they are added to revenue in a profit and loss account. Opening valuation is included in the deduction stage for the same reason that it is added to expenditure in a profit and loss account.

Follow the calculation through and check each figure. Debtors, creditors and valuations have been kept deliberately simple for the purpose of this exercise.

The basic principles followed apply for any type of business. If your interest is horticulture, manufacture or commerce you will find that direct comparisons may be made at each stage; it is all business after all, whatever commodity is being produced or sold.

Calculation of Trading Profit	£
Trading net cash flow annual total	32,362
Notional receipts (produce consumed, house rent, private use of car etc.)	
Closing valuation – livestock	58,000
– harvested crops	42,450
– tillages and growing crops	
– stores	
Closing debtors	7,350
Opening creditors	1,500
Sub Total (a)	141,662
Opening valuation – livestock	58,700
– harvested crops	37,450
– tillages and growing crops	
– stores	
Closing creditors	750
Opening debtors	6,800
Sub Total (b)	103,700
Profit (a–b) before depreciation	37,962
Depreciation on buildings and fixtures	
Machinery depreciation	10,993
Trading profit after depreciation	26,969

MONITORING CASH FLOW

Now we have the budget cash flow, what do we do with it? Well, a great deal of effort is wasted if the good work is not continued.

Remember, a budget cash flow shows likely peak borrowing requirements, enabling arrangements to be made in advance. In addition to this, performance targets were set in the construction of the cash flow, so enabling comparison of budget cash flow with actual cash flow.

As the business proceeds, cash book records are compiled and period totals are transferred to 'Actual' columns on the cash flow for comparison with budget. It is important that cash book column headings are in agreement with cash flow detail so that monthly or quarterly totals may be transferred with minimum difficulty.

Individual actual receipts and payments are compared with budget figures and differences noted as increase or decrease. A separate sheet or form is best used for this purpose, with the same narrative as the cash flow form. This avoids cluttering the cash flow and a fresh monitoring sheet can be used for each month, quarter or other period.

Difference or variance between budget and actual cash flow may have a simple explanation which is easily determined – that accident repair job or increase in the rent. Many variances may be relatively insignificant. Major influences on cash flow over which there is management control must be looked at more closely and the variance analysed to allow corrective action to be taken if necessary.

VARIANCE ANALYSIS

Variance between budget and actual cash flow can always be identified with one or more of three possible reasons as follows.

Variance from plan The physical plan has not been followed; more or less productive units have been used – more or less area, breeding livestock or other unit of production.

Variance from yield or quantity The yield, output or input per unit of production varies from that predicted.

Variance from price The unit price of goods bought or sold varies from that predicted.

In each case there may be one or more underlying reason, some controllable by management and some not under that control. Poor technical management may lead to disease and lowered production, inflation may cause price rises.

Reasons for variance must be identified and corrective action taken where possible. By this means cash flow is kept under control and lenders of finance will have confidence in management.

GREEN FARM – VARIANCE ANALYSIS OCT./NOV./DEC.

Actual cash flow for the first three months is compared with budget, variance of key receipts and payments analysed and a report written. The report is simply a statement of fact, it is then up to management to consider further underlying reasons, such as those given above, and decide on what action is to be taken.

Calculating variance

To determine whether variance is + or – in effect, ask the question "Does the variance improve cash flow or does it make it worse?". If receipts are greater they are positive, if less they are negative. If payments are greater they are negative and if less they are positive, or an improvement.

Note in the example that the sum of the variance is equal to the difference between budget and actual totals at each stage.

TRADING

	Budget		Actual		Variance
Receipts	Quantity	£	Quantity	£	£
Milk	90,500lt	18,100	85,121lt	17,450	–650
Calves	49	6,440	45	6,075	–365
Cull cows	13	5,850	15	6,900	1,050
Barren heifers					0
Wheat (19X9 Crop)	200t	20,600	203t	19,691	–909
Barley (19X9 Crop)	30t	2,850	29t	2,755	–95
Sub Total (a)		53,840		52,871	–969

TRADING	Budget		Actual		Variance
Payments	Quantity	£	Quantity	£	£
Dairy cow feed	40 t	5,980	39 t	5,850	130
Dairy youngstock feed	6 t	770	6 t	800	−30
Vet. & med.		600		576	24
Livestock sundries		750		690	60
Seed		2,800		2,950	−150
Fertiliser		3,500		3,400	100
Crop protection		2,200		2,350	−150
Fuel,oil,elec		1,100		1,250	−150
Machinery costs		1,500		1,150	350
Contract		400		425	−25
Wages		5,200		5,355	−155
General repairs		800		620	180
Rent, rates, water		3,800		3,850	−50
Gen. ins., sundry fixed costs		400		420	−20
Admin., telephone		800		760	40
Sub Total (b)		30,600		30,446	154
Net cash flow (a−b)		23,240		22,425	−815
Interest		(1,485)		(1,485)	
Trading net cash flow (A)		21,755		20,940	
PERSONAL					
Receipts, sub total (c)					
Drawings		2,500		2,800	
Tax					
Any other payments					
Sub Total (d)		2,500		2,800	300
Personal cash flow (B)		(2,500)		(2,800)	−300
CAPITAL					
Grants					
Mach. sales					
Sub Total (e)					
Buildings etc.					
Machinery	Tractor	10,000		10,400	400
Other capital repayments					
Sub Total (f)		10,000		10,400	400
Capital cash flow (C) (e−f)		(10,000)		(10,400)	
Net cash flow (A+B+C)		9,255		7,740	−1,515
Cumulative balance		(23,745)		(25,260)	−1,515
Opening balance		(33,000)			

Calculation of variance analysis – Green Farm

As stated earlier, variance will be due to change in plan, yield/quantity and/or price. Calculation of the contribution which one or more of these changes make to variance from budget is as follows.

Variance from plan (i.e. units of production) The effect of variance from plan is calculated by multiplying that variance by the budget yield/quantity and budget price: variance in plan × budget yield/quantity × budget price.

Variance from yield quantity The effect of this variance is calculated by multiplying the variance by actual numbers of units of production (area of land, numbers of productive livestock, etc.) and budgeted price: variance in yield/quantity × actual units of production × budget price.

Variance from price This is calculated by multiplying variance in price by actual units of production and actual yield/quantity: variance in price × actual units of production × actual yield/quantity.

INTERPRETING THE ANALYSIS

Analysis of one item of receipt variance is shown below, following the formula given earlier.

Milk

Receipts are down £650 compared with budget; we need to establish the reason for this. From the livestock numbers record (not shown), it is found that there were only 81 cows compared to the budgeted 84.

	Cows	Litres/Cow	Pence/Litre
Budget	84	1,077.3	20.0
Actual	81	1,050.8	20.5
Variance	−3	−26.5	+0.5

Variance from budget due to:

		£ Variance + −
Plan	3 × 1,077 × 20.0p	646−
Yield	26.5 × 81 × 20.0p	429−
Price	0.5p × 81 × 1,050	425+
Total milk variance		650−

From the variance analysis we find that failure to keep to plan with regard to cow numbers and yield has caused a fall in milk receipts of £1,075 (646 + 429). This has been partially offset by an improvement in milk price. A common sense approach should be taken with regard to selecting items for variance analysis and information gained should be put to good use in rectifying weaknesses. Further analysis may be completed from the detail provided.

CHAPTER 15

More from the accounts

Having considered current enterprise performance by means of gross margin analysis, produced a plan, budget and cash flow for the coming year it is time to take a broader look at the business. Indeed, it could be argued that this should have been done before short term performance analysis. In fact, the strengths and weaknesses of any business must be kept under constant review.

A business can be compared to a living (or dying) body on which the passage of time and all the surrounding influences leave their mark for good or ill. The main difference between a business and a natural body is that a business body does not necessarily grow weak and decrepit with age; but it does happen and sometimes prematurely too!

A key word which must be uppermost in the mind when considering the state of a business is **trend.** Comparison may aptly be made once more with a natural body. If signs of health and wellbeing are improving from a low state – wonderful! If there is decline from good health – anxiety and visions of doom. Any and every business should be given a health check on a regular basis to determine whether it is on the up or down – it is usually one or the other. A state of equilibrium rarely exists.

A bank manager's first interest and means of assessing the state of a business is the balance sheet, from which he may extract certain informative detail (more to follow on this). But the manager may wish to visit the business premises, before paying too much attention to the balance sheet, and use his eyes and common sense. He will make mental notes on the general state of premises and equipment and the level of staffing. If an established business has a rather run down appearance, has old and decrepit equipment and is under-staffed, then first signs are that 'room to manoeuvre' by making economies is limited and vice versa.

The next stage in weighing up the business will require a look at the balance sheet(s) to determine total borrowing and a look at the profit and loss account(s) to determine total turnover (value of trading sales). Trends of these two figures over several years will either encourage or alarm. If turnover is increasing and borrowing falling then in very broad terms that is good news. If turnover is falling and borrowing is increasing, then clearly there is a problem that must receive urgent attention.

The point may be suitably made here that profit, which may appear good on its own, is of little value in determining the strength of a business. A run of apparently successful years in terms of profit gained may mask a situation of decline and failure.

Balance sheet ratios build the first picture of a business. Many farm managers and owners have difficulty reading and interpreting a balance sheet and hence have little idea of what a telling picture it gives to others. We now consider the most important measures or ratios obtained from a balance sheet. When they are compared with the same measurements for previous years then a trend may be determined, for good or ill.

CAPITAL OR EQUITY

Equity is a term which may be used to indicate the owner's (or more than one owners') stake in the business relative to loan capital. Equity may also be defined as 'the owner's stake in the business expressed as a percentage of total assets'. Some go further in clarifying the interpretation or use of the term by modifying it to 'percentage equity'. The important thing is to know where to look for the information and to understand its significance.

Equity and total assets

If the total assets of a business are worth £100,000 and the owner's share is £60,000 then the first calculation tells us that the proprietor owns more than half of the total investment and is therefore fully in control (hopefully, anyway!). A further calculation, easy in this example, tells us that the proprietor owns 60 per cent of total assets.

$$\frac{60{,}000}{100{,}000} \times 100 = 60\% \text{ equity}$$

This is a reasonably secure position for the proprietor, but a decrease in equity below this level should cause the alarm bells to ring. Ideally the general trend over the years of trading should indicate an increase in equity – but this must be looked at in the context of other activities in which the owner is involved. It could be that borrowing has increased in order to release capital for some other purpose, such as assisting a son or daughter to get into business.

LIQUIDITY

Having established the overall position of the proprietor, the financial structure of the business must be considered from several other viewpoints. Of immediate interest is the ability of the business to pay creditors without selling capital assets such as productive equipment or productive livestock. Two balance sheet ratios are calculated to indicate this level of liquidity.

Ability to pay – liquidity ratios

The two measures of ability to pay, expressed as a ratio, calculate the ability to meet short term liabilities (normally meaning creditors) from current assets. The first measures the ability to pay by reference to all of the current assets, the second bases the ability to pay on liquid assets, which include only debtors and bank balance.

Current ratio

$$\frac{\text{current assets}}{\text{current liabilities}} = \text{current ratio}$$

If a balance sheet shows current assets to be worth £30,000 and current liabilities to be £20,000 then

$$\frac{30{,}000}{20{,}000} = \text{current ratio } 1.5 : 1$$

This indicates that there is enough value in current assets to meet current liabilities one and a half times over if suddenly called upon to do so. This is a reasonably sound position to be in, but it may not be helpful to have to sell sale stock prematurely, and so the following measure must also be made.

'Acid test' ratio

$$\frac{\text{current assets} - \text{stock}}{\text{current liabilities}} = \text{'acid test' ratio}$$

This measurement determines the ability to pay creditors without disposing of stock – in other words from the bank account and from funds owed by debtors.

If we refer to the balance sheet again and find that the value of stock is £10,000, we calculate as follows:

$$\frac{30{,}000 - 10{,}000}{20{,}000} = \text{acid test ratio } 1 : 1$$

This result indicates that creditors can be paid in the short term without disposing of further assets (providing of course that debtors pay up).

Liquidity ratios – bank overdraft

If the balance sheet shows an overdraft at the bank, the overdraft should be included in the current liabilities if meaningful values are to be calculated. It must be recognized that an overdraft is always considered short term and can be called in by the bank at short notice.

Meaning of 'stock' in liquidity ratios

The amount and type of stock held by different types of business will vary considerably. The methods given for calculating liquidity ratios apply to businesses in general. It may be considered more sensible to include only stock which is intended for sale, rather than all current assets in the case of agriculture, due to the lengthy production cycle of most farm products. Consistency of method is essential; so once having determined the approach, stick to it.

GEARING OF LOAN CAPITAL TO EQUITY

Having established the position with regard to equity and ability to pay, the relationship between loan capital and owner's capital must be considered. Equity expressed as a percentage has already given a clue to this.

Long and medium term liabilities are effectively an investment by 'outsiders' in the business. It is important to establish the proportion of investment by outsiders to that by the proprietor and this is expressed as *gearing ratio* or *gearing percentage*.

A low level of borrowed capital relative to owner's capital is known as low gearing. A high level is known as high gearing – which also implies high pressure on the business to meet interest charges.

Gearing Ratio Loan capital: Equity (as a value of 1)

$$\frac{\text{Loan capital}}{\text{Owner's capital}} \quad \frac{£10000}{£100000} \quad \frac{0.1}{1.0} \qquad \text{Ratio } 0.1:1 \qquad \text{Low gearing}$$

$$\frac{\text{Loan capital}}{\text{Owner's capital}} \quad \frac{£100000}{£100000} \quad \frac{1.0}{1.0} \qquad \text{Ratio } 1:1 \qquad \text{High gearing}$$

Gearing Percentage Loan capital as % of (equity + loan capital)

$$\frac{\text{Loan capital}}{\text{Capital employed}} \quad \frac{£10000}{£110000} \times 100 = 9\% \qquad \text{Low gearing}$$

$$\frac{\text{Loan capital}}{\text{Capital employed}} \quad \frac{£100000}{£200000} \times 100 = 50\% \qquad \text{High gearing}$$

Ratio and percentage tell the same story but percentage gives the more instant picture and is more widely used.
N.B. Loan capital = (medium + long term loans). Consistency is *essential* to accuracy when calculating periodic trends.

SOLVENCY AND INSOLVENCY

Equity represents the owner's stake in the business. Whilst equity exists, then all liabilities of the business can be met if necessary even if the business has to be sold, but it is **solvent**.

When there is no equity, that is when liabilities are greater than assets, the business is **insolvent** since it is unable to meet its debts in full (it cannot dissolve them) even if the business is sold. Under these circumstances the business is likely to become bankrupt.

RETURN ON CAPITAL

Having established that the business is in reasonable shape we look now for an indication of overall performance by expressing profit as a percentage return on capital. The big difficulty here is that there are many ways of defining capital and the amounts can vary considerably. To mention a few measures of capital, there is **initial capital**, being the capital invested when business commenced or changed direction, there is **average capital, tenant's capital** and **landlord's capital**. What we are not considering is equity or owner's capital investment.

There is a further complication in that business premises differ. For some it may be reasonable to include capital invested in premises in the calculation and for others it may be quite ridiculous. It has long been impossible to gain a realistic return on capital invested in farm land by farming; this may or may not be gained only through increase in capital value.

The most commonly used measure of capital is **average tenant's capital** which includes stock, machinery and equipment. The average is determined from opening and closing values for the year.

$$\frac{\text{profits for the year}}{\text{average tenants capital}} \quad \frac{20{,}000}{100{,}000} \times 100 = 20\% \text{ p.a.}$$

The calculation shows a 20 per cent return on capital: is this sufficiently attractive? This of course depends on current bank interest rate, if there is a 4 or 5 per cent margin over the cost of bank interest then it is worthwhile in business terms. Who, though, goes into business purely for profit? There are many other aspects to consider, such as lifestyle and job satisfaction, to which it is difficult to attach a monetary value.

NET FARM INCOME AND MANAGEMENT AND INVESTMENT INCOME

Farmers have the benefit of published information giving average results of farms in particular categories and areas. This work is carried out by the government sponsored Farm Management Survey (FMS). Regional centres which are normally attached to the agricultural economics departments of a university conduct annual surveys on selected farms in their area.

The published results enable individual farmers to compare their own results with others in the same category, but a number of adjustments have to be made to ensure that it is a fair comparison.

Adjustments to profit

To give net farm income (NFI):

a) value of unpaid family labour (excluding farmer and wife) is added to expenditure;
b) interest charges are deducted from expenditure;
c) non-farm revenue and expenditure are deducted;
d) ownership expenses are deducted from expenditure;
e) a notional rent is added to expenditure.

Determination of NFI gives a reasonable level of standardisation to the account for comparison purposes. The next stage is to deduct from NFI the estimated value of the farmer and wife's labour (or farmer and husband/partner as the case may be) to determine profit gained as a result of management skills and investment of capital, this being known as **management and investment income** (MII).

Clearly, having calculated MII it makes sense to calculate return on capital. To take the simple example used in the previous calculation of return on capital, after adjustments the NFI was calculated at £25,000.

	£
NFI	25,000
Less farmer/wife's labour	6,000
MII	19,000

(He may only work physically on a part-time basis.)

Return on capital

$$\frac{\text{MII}}{\text{Average tenant's capital}} \quad \frac{£19{,}000}{100{,}000} = 19\% \text{ p.a.}$$

This is only part of the comparison with similar farms; further detail may be gained from university publications or regional FMS and from specialist farm management textbooks.

TIME TAKEN TO PAY

Businesses which make regular supplies on a credit basis to a number of customers are likely to face cash flow problems due to slow settlement of debts. Obviously the first line of attack is to maintain up-to-date records and to pursue late payers for the money – but what is the overall situation, is it under control or is it getting worse? One means of monitoring the situation is calculation of average time taken by debtors before settlement of their accounts.

First, calculate average sales per day

$$\frac{\text{annual sales}}{365} = \frac{287{,}000}{365} = \text{£786 per day}$$

Average number of days credit taken by debtors

$$\frac{\text{debtors}}{\text{average sales per day}} = \frac{28{,}306}{786} = \text{36 days}$$

This result tells us that the average debtor is taking more than normal trade terms of 28 days and so it is clear that when prompt payers are allowed for, some are really taking advantage and must be chased up. Calculation of this measure on a regular basis enables **trend** to be determined.

CHAPTER 16

Ratios and trends for Green Farm

In this chapter we take another look at Green Farm and apply the measures discussed earlier. 'Time to pay' is not included since it is not relevant to this type of business. For ease of reference, the two balance sheets and capital accounts which appeared in Chapter 7 are reproduced, but in a different format this time. The reader should compare and check content.

More than two years' accounts and calculated ratios are required to indicate a clear trend but even so, movement from one year end to another provides useful indicators of what is happening. Strengthening of position over a year must be encouraging whilst adverse results should prompt a closer watch on the business.

Calculation of such information has added value in that not only is information gained on the state of the business and the direction in which it is travelling, but greater credibility is gained with the bank manager from whom support may be required. There is also considerable satisfaction in gaining such precise information rather than remaining in ignorance of the true state of the business.

Balance Sheet for Green Farm

	as at 31.3X8		as at 31.3.X9	
Fixed assets	£	£	£	£
Machinery	44,720		38,206	
Fixed equipment	912		760	
Dairy cows	45,000		39,000	
Total fixed assets		90,632		77,966
Current assets				
Trading livestock	13,300		13,700	
Harvested crops	475		105	
Stores	10,627		12,136	
Bank	–		–	
Debtors	9,867		13,671	
Total current assets		34,269		39,612
Total assets		124,901		117,578

Liabilities	£	£	£	£
Medium/long term loan	7,614		2,614	
Capital (owner's)	79,724		82,180	
Total medium/long term		87,338		84,794
Current liabilities				
Bank	30,120		23,989	
Creditors	7,443		8,795	
Total current liabilities		37,563		32,784
Total liabilities		124,901		117,578

Capital at 31.3.X7	67,782		31.3.X8	79,724	
Profit for year	26,186			17,797	
		93,968			97,521
Drawings		14,244		(20,341) (–5,000)	15,341
Capital at 31.3.X8		79,724	31.3.X9		82,180

CALCULATION OF ACCOUNTING RATIOS – GREEN FARM

Equity, expressed as a percentage of total assets.

at 31.3.X8

$$\frac{79{,}724}{124{,}901} \times 100 = 63.83\%$$

at 31.3.X9

$$\frac{82{,}180}{117{,}578} \times 100 = 69.89\%$$

There is a definite improvement here but there are several fairly obvious reasons for this. There has been a modest introduction of (£5,000) funds and the cows have been carefully but rigorously culled, so reducing capital invested.

LIQUIDITY

Current ratio is considered first. The problem arises as to which of the current assets to include in this calculation. It would be a bit pointless to include purchased stocks of 'raw materials' such as fertiliser in this calculation because they are not available for disposal – unless the business winds up. For this example we include trading livestock, harvested crops and debtors.

at 31.3.X8

$$\frac{23{,}642}{37{,}563 \text{ (inc. overdraft)}} = $$ current ratio 0.6 : 1 (i.e. only 60p is available to meet each £1 owed)

at 31.3.X9

$$\frac{27{,}476}{32{,}784}$$ = current ratio 0.8 : 1

Whilst there is a definite improvement in the second year, neither situation is desirable. If the overdraft were called in at short notice an embarrassing situation could arise – perhaps the overdraft should be converted to a loan.

If the overdraft is excluded from the calculation we see a much stronger ability to pay and quite honestly with the equity as strong as it is there is little likelihood of overdraft facility being reduced.

at 31.3.X8

$$\frac{23{,}642}{7{,}443}$$ = current ratio 3.0 : 1

at 31.3.X9

$$\frac{27{,}476}{8{,}795}$$ = current ratio 3.12 : 1

ACID TEST RATIO

This ratio really does indicate ability to pay debts in the short term, but again the question of the bank overdraft has to be considered.

at 31.3.X8

$$\frac{9{,}867}{37{,}563}$$ = acid test ratio 0.3 : 1 30p available for each £1 owed

or excluding bank overdraft

$$\frac{9{,}867}{7{,}443}$$ = acid test ratio 1.3:1 £1.30 available for each £1 owed

at 31.3.X9

$$\frac{13{,}671}{32{,}784}$$ = acid test ratio 0.4:1

excluding overdraft

$$\frac{13{,}671}{8{,}795}$$ = acid test ratio 1.6:1

Again we see an improvement in the second year.

What exactly does the ratio signify? The '1' represents £1 owing and the first figure shows the ability to pay that pound. In the first of the acid test results, only 30p is available to pay each £1 of debt and in the final calculation £1.60 is available to pay each £1 of debt.

GEARING RATIO

In this case we consider the ratio of loan capital to equity and suggest that the reader calculates gearing percentage and compares the clarity of information.

at 31.3.X8

$$\frac{7{,}614}{79{,}724} = \text{ratio } 0.1{:}1 \quad \text{(10p borrowed: £1 owner's capital)}$$

at 31.3.X9

$$\frac{2{,}614}{82{,}180} = \text{ratio } 0.03{:}1$$

The medium/long-term borrowed capital is really very low. Suppose a £20,000 loan is arranged in the second year as an alternative to overdraft at the bank:

$$\frac{22{,}614}{82{,}180} = \text{ratio } 0.28{:}1 \quad \text{(still a low gearing)}$$

In this final situation gearing is still low with only 28 pence borrowed against each £1 of owner's capital. There is scope for further borrowing for expansion or diversification. Calculation of year-end gearing will provide a pointer to the progress, strength and stability of the business. Now calculate gearing percentage.

RETURN ON CAPITAL

The vexed question of what is considered capital for this purpose has already been considered. Average of tenant's capital is taken in this instance and this is probably the most commonly used for businesses of this type. The more widely used measure, generally applied to non-farming businesses, is also calculated.

Referring again to the two balance sheets and capital accounts, information is extracted to calculate tenant's capital, this being the capital required to operate the business as against owning the premises. Only two years of full information is provided on the balance sheet, the total for the previous year is provided here to enable average capital to be calculated for each of the two years.

TENANT'S CAPITAL – FROM BALANCE SHEETS

	31.3.X7	31.3.X8	31.3.X9
	£	£	£
Machinery		44,720	38,206
Fixed equipment		912	760
Dairy cows		45,000	39,000
Trading livestock		13,300	13,700
Harvested crops		475	105
Stores		10,627	12,136
	112,650	115,034	103,907

Return on tenant's capital

31.3.X8
Average capital

$$£\frac{112{,}650 + 115{,}034}{2} = £113{,}842$$

$$\frac{\text{Profit}}{\text{Av. capital}} \quad \frac{26{,}186}{113{,}842} \times 100 = 23\%$$

31.3.X9

Average capital

$$£\frac{115{,}034 + 103{,}907}{2} = £109{,}470$$

$$\frac{\text{Profit}}{\text{Av. capital}} \quad \frac{17{,}797}{109{,}470} \times 100 = 16.25\%$$

When comparison between farms is being made then adjustment is made to the profit figure. See notes on FMS, Net Farm Income and Management and Investment Income.

Return on owner's capital (at start of business year)

This may be seen as a more sensible measure since it indicates the return gained by the owner's investment. But there are arguments for and against this line of thought. The important thing is to select those measures which are considered applicable to the business type under consideration.

31.3.X8

$$\frac{\text{Profit}}{\text{Owner's capital}} \quad \frac{26{,}186}{67{,}782} \times 100 = 38.6\%$$

31.3.X9

$$\frac{\text{Profit}}{\text{Owner's capital}} \quad \frac{17{,}797}{79{,}724} \times 100 = 22.32\%$$

There is a quite dramatic fall in return in the second year, reflected in both return on tenant's capital and owner's capital. Clearly further detailed investigation is required to determine the reason for this and to put it right.

CHAPTER 17

A step further with double entry book-keeping for a horticultural business

In this chapter we give further consideration to double entry book-keeping. A horticultural business is used for the example. Leaffe Brothers are growers who sell their produce wholesale to local garden centres and retail through their own shop. Some goods such as compost are purchased for resale. For the purposes of double-entry book-keeping, we have to remember that no business activity can take place unless at least two people are involved. On each occasion the two people must each record their side of the transaction. We consider a few examples.

A customer purchases goods on credit

The sales department has provided the customer with goods so an entry is made on the 'giving' side of their sales book; it is a **credit** entry. The customer has received goods and so must enter detail on the 'receiving' side, showing indebtedness to the business as a **debit** in his **personal ledger**.

Sales has given or credited, customer has received and so is indebted.

The customer 'gives' money and so a credit entry is made in his personal ledger. Bank receives the money and so is indebted to the customer; entry is made on the debit side of the bank ledger.

Likewise, when the purchaser buys goods for the business, the goods are received and the purchaser is indebted: a debit entry is made. The supplier has given the goods and so is credited: a credit entry is made in his books.

Who gives and who receives

Think through who gives; separate giving and receiving of goods from giving and receiving of money. When a customer pays for goods at the time of purchase, giving is balanced by receiving and so may not

be entered in his personal account but, in this case, bank receives money (debit) and sales give goods (credit). Get used to this process and then the names of the different types of account will become obvious.

Trial balance

At the end of a series of transactions total credits must equal total debits and this represents the **trial balance**. A trial balance may be drawn up at any point, and when both credit and debit appear in a ledger, then only the balance is shown in the trial balance.

We look at our horticultural business towards the end of a year's trading. Books have been balanced monthly and trial balances produced. The trial balance given below shows the position at the end of eleven months and individual accounts should be checked against this to ensure you understand the source of information. The business year ends on 30th September.

You will note that stock, capital and plant and machinery values are given as at the start of the business year. This is normal practice; the values are updated at the year end with stocktaking, calculation of machinery and equipment depreciation and determination of capital when the balance sheet is drawn up.

Purchase and sales totals have been accumulated through the year. This is preferably done on a summary sheet similar to that given in the cash analysis exercise but not shown here in full.

Personal accounts are limited in number for simplicity but principles are still demonstrated.

Trial Balance at 31 August 19XX

	Dr £	Cr £
Stock at start of year	38,450	
Capital at start of year		122,034
Bank		4,200
Cash in hand	450	
Loan		37,300
Purchases	102,427	
Sales		154,536
Bank charges and interest	5,870	
Debtors	5,396	
Creditors		2,743
Opening valuation of:		
plant and machinery	28,550	
vehicles	16,270	
buildings and glass	123,400	
	320,813	320,813

Why does the trial balance always balance (well, after finding and correcting errors)? Remember, for every debit (Dr) entry there is a credit (Cr) entry. When goods are sold (Cr), debtors or bank will be Dr. Likewise when goods are purchased (Dr), creditors or bank will be Cr. This occurs with each transaction, Dr and Cr, throughout the trading period.

Next, we examine the purchase and sales accounts, the nominal accounts and the real accounts. The nominal and real accounts show balances brought forward to 1st September. The purchase and sales accounts show September transactions, and then the totals for the first eleven months are brought forward and added to the total for that month. Entries are made throughout the final month; refer to the list of transactions and check each debit and credit entry.

List of transactions for September

3rd Sold shrubs to Sunny Garden Centre on credit £500
5th Purchased trays from Plastic Mouldings PLC on credit £200
6th Cash sales bedding plants £150, compost £40, pots £65
7th Sold fruit trees £1,200, ornamental trees £800 to Newvale Garden Centre on credit
7th Banked £500 cash
8th Paid insurance on van £400
14th Sold bedding plants to Sunny Garden Centre on credit £1,400
15th Purchased on credit, boiler fuel £1,200 from Hortic. Supplies Ltd
15th Cash sales, bedding plants £370, compost £28
17th Sold to Newvale Garden Centre on credit, ground cover £263, house plants £120, roses £570, climbers £125
17th Banked £250 cash
20th Cheque received from Sunny Garden Centre £1,000 in part settlement of account
23rd Purchased liners on credit from Cuttings & Seeds Ltd £180
23rd Cash sales, ground cover £85, house plants £126, shrubs £83, bedding plants £150
26th Paid petrol bill for month on receipt of invoice £260
28th Paid Hortic. Supplies £860, Plastic Mouldings £960, Cuttings & Seeds £637, Wages 3,400
28th Cheque received from Newvale Garden Centre £3,000

VAT has been omitted from the transactions in order to concentrate on basic principles of double entry book-keeping. A section on pages 160–161 considers methods of accounting for VAT.

Coding of accounts for cross referencing has not been included in this example. See Chapter 4, section on design of book-keeping stationery, for method. Lightly pencil in suitable cross reference codes as you check entries.

Figure 17.1 Purchase and sales account for Leaffe Bros

Cr — SALES — Year Ending 30. 9. xx

S	Date	Detail	TOTAL	Cheque No.	FRUIT TREES BUSHES	ORN. TREES BUSHES	SHRUBS	ROSES	GROUND COVER	CLIMBING	BEDDING	HOUSE PLANTS
	SEPT				£ p.	£ p.	£ p.	£ p.	£ p.	£ p.	£ p.	£ p.
1	3	Sunny G. Centre.	500 -				500 -					
2	6	Cash sales	255 -								150 -	
3	7	Newvale G. Centre.	2000 -		1200	800						
4	14	Sunny G. Centre.	1400 -								1400 -	
5	15	Cash sales	398 -								370 -	
6	17	Newvale G. Centre	1078 -					570 -	263 -	125 -		120 -
7	23	Cash sales	444 -				83 -		85 -		150 -	126 -
8		TOTAL SEPT.	6075		1200	800	583	570	348	125	2070	246
9		11 months b.f.	154536		6060	14632	16251	12490	2973	4484	86230	6923
10		TOTAL SALES	160611		7260	15432	16834	13060	3321	4609	88300	7164
11												
12												
13												

Dr — PURCHASES — Year Ending 30. 9. xx

P	Date	Detail	TOTAL	Cheque No.	SEED	LINERS	FERTILISER	CROP PROTECTION	COMPOST	TRAYS POTS	STAKES	PLASTIC SHEET
	SEPT				£ p.	£ p.	£ p.	£ p.	£ p.	£ p.	£ p.	£ p.
1	5	Plastic Mouldings. Trays	200 -							200 -		
2	8	Nor. Union. Van ins.	400 -									
3	15	Cuttings + Seeds. Liners	180 -			180 -						
4	23	Hortic. Supplies. Fuel.	1200 -									
5	26	Petrol.	260									
6	28	Wages	3400									
7		TOTAL SEPT	5640		–	180	–	–	–	200 -	–	–
8		11 months b.f.	102427		3250	2700	1650	2480	8560	9725	1160	1570
9		TOTAL PURCHASES	108067		3250	2880	1650	2480	8560	9925	1160	1570
10												
11												
12												

Personal Accounts

Horticultural Supplies Ltd

Dr		£	Cr		£
28 Sept.	Bank	860	Bal. b.f.	1 Sept.	860
Cr bal. c.f.		1,200	15 Sept.	Boiler fuel	1,200
		2,060			2,060

Plastic Mouldings Ltd

Dr		£	Cr		£
28 Sept.	Bank	950	Bal. b.f.	1 Sept.	1,246
Cr bal. c.f.		496	5 Sept.	Trays	200
		1,446			1,446

Cuttings & Seeds & Co

Dr		£	Cr		£
28 Sept.	Bank	637	Bal. b.f.	1 Sept.	637
Cr bal. c.f.		180	23 Sept.	Liners	180
		817			817

SALES

Year Ending 30. 9. XX

Line	COMPOST £	p.	POTS. TRAYS SUNDRY £	p.	PREPACK FERT. £	p.
1						
2	40	-	65	-		
3						
4						
5	28	-				
6						
7						
8	68		65			
9	2130		1623		740	
10	2198		1688		740	
11						
12						
13						

PURCHASES

Year Ending 30. 9. XX

Line	SUNDRY DIRECT £	BOILER FUEL £	ELECTRICITY £	WATER £	ADMIN. PHONE. INS £	VEHICLE COSTS £	RENT RATES £	WAGES £
1	–							
2	–					400 -		
3	–							
4	–	1200 -						
5	–					260 -		
6	–							3400 -
7	–	1200	–	–	–	660		3400
8	930	12600	3870	2465	3224	4850	6653	36240
9	930	13800	3870	2965	3224	5510	6653	39640
10								
11								
12								

Personal Accounts (cont.)

Sunny Garden Centre

Dr		£	Cr		£
Bal. b.f.	1 Sept.	2,120	20 Sept.	Bank	1,000
3 Sept.	Shrubs	500			
14 Sept.	Bedding plants	1,400	Dr bal. c.f.		3,020
		4,020			4,020

Newvale Garden Centre

Dr		£	Cr		£
Bal. b.f.	1 Sept.	3,276	28 Sept.	Bank	3,000
7 Sept.	Trees	2,000			
17 Sept.	Sundry	1,078	Dr bal. c.f.		3,354
		6,354			6,354

Real Accounts

Stock

Dr	£	Cr
Opening bal.	38,450	

Plant & Machinery

Dr	£	Cr
Opening bal.	28,550	

Vehicles

Dr	£	Cr
Opening bal.	16,270	

Buildings & Glass

Dr	£	Cr
Opening bal.	123,400	

Bank Account

Dr		£	Cr		£
7 Sept.	Cash	500	Bal. b.f.	1 Sept. 19XX	4,200
17 Sept.	Cash	250	8 Sept.	N. Union	400
20 Sept.	Sunny Vale	1,000	26 Sept.	Petrol, Fosse Garage	260
28 Sept.	New Vale	3,000	28 Sept.	Hortic. Supplies	860
			28 Sept.	Plastic Moulds	950
		4,750	28 Sept.	Cuttings & Seeds	637
Cr. bal. c.f.		5,957	28 Sept.	Wages	3,400
		10,707			10,707

Cash Account

Dr		£	Cr		£
Bal. b.f.	1 Sept.	450	7 Sept.	Bank	500
6 Sept.	Cash sales	255	17 Sept.	Bank	250
15 Sept.	Cash sales	398			
					750
23 Sept.	Cash sales	444	Dr bal. c.f.		797
		1,547			1,547

Loan

Dr	£	Cr		£
		Bal. b.f.	1 Sept. 19XX	37,300

Interest

Dr	£	Cr	£
Bal. b.f.	5,870		

Trial balance at 30 September 19XX	Dr	Cr
	£	£
Stock at start of year	38,450	
Capital at start of year		122,034
Bank		5,957
Cash in hand	797	
Loan		37,300
Purchases	108,067	
Sales		160,611
Bank charges and interest	5,870	
Debtors	6,374	
Creditors		1,876
Opening valuation of		
Plant and machinery	28,550	
Vehicles	16,270	
Buildings & glass	123,400	
	327,778	327,778

Note:

Total debtors taken from personal accounts	£
Sunny Garden Centre	3,020
Newvale Garden Centre	3,354
	6,374

Total creditors taken from personal accounts	£
Hortic. Supplies Ltd	1,200
Plastic Mouldings	496
Cuttings & Seeds	180
	1,876

The final month of trading is complete and a trial balance has ensured that the books are all in order. The next stage is to prepare year end accounts.

YEAR END PROCEDURE

Several tasks must be completed before final accounts are produced, these include **stocktaking** and calculation of **depreciation**. Profit is calculated in two stages: **gross profit** is determined in the trading account, and **net profit** in the profit and loss account. The profit and loss account could equally provide another learning stage by considering the two part accounts.

Stocktaking

The stock account carries the value of opening stock throughout the trading period and this is not changed until final accounts are drawn up and then the new value enters the books as the opening figure for the new year. Opening and closing value are treated in the same way as in earlier accounts in principle, but with a slightly different method. Justification of the procedure in purist double entry terms can be made, but is a little obscure and probably unimportant at this level.

Stock Account
(Year 1)

Dr		£	Cr		£
Oct. 1	Opening bal.	38,450	Sept. 30	Transfer to trading a/c	38,450
Oct. 1	(Year 2) Op. bal.	36, 245			

Depreciation account

No capital purchases have been made this year so calculation is simple. Plant and machinery and vehicles are depreciated at 20 per cent of written down value buildings and glasshouses at 10 per cent of cost. Full background detail of cost and earlier depreciation is omitted in the interest of simplicity.

Depreciation account

Plant & Machinery
(Year 1)

Dr	£	Cr		£
Opening bal. W.D.V.	28,550	30 Sept.	Depreciation to p&l a/c	5,710
		30 Sept.	bal. c.f.	22,840
	28,550			28,550
(Year 2)				
Opening bal. W.D.V.	22,840			

Vehicles
(Year 1)

Dr	£	Cr		£
Opening bal. W.D.V.	16,270	30 Sept.	Depreciation to p&l a/c	3,254
		30 Sept.	bal. c.f.	13,016
	16,270			16,270
(Year 2)				
Opening bal. W.D.V.	13,016			

Buildings & Glasshouses
(Year 1)

Dr	£	Cr		£
Opening bal. W.D.V.	123,400	30 Sept.	Depreciation to p&l a/c	14,000
		30 Sept.	bal. c.f.	109,400
	123,400			123,400
(Year 2)				
Opening bal. W.D.V.	109,400			

Trading and profit and loss account for B & R Leaffe for year ending 30th September 19XX

Trading Account

	£		
Opening Stock	38,450		
Purchases		*Sales*	
Seed	3,250	Fruit trees	7,260
Liners	2,880	Orn. trees	15,432
Fertiliser	1,650	Shrubs	16,834
Crop protection	2,480	Roses	13,060
Compost	8,560	Ground cover	3,321
Trays, pots	9,925	Climbing plants	4,609
Stakes	1,160	Bedding plants	88,300
Plastic sheet	1,570	House plants	7,169
Sundry direct costs	930	Compost	2,198
		Pots, trays	1,688
		Prepack ferts.	740
	70,855		
Less closing stock	36,245		
Cost of stock sold	34,610		
Gross profit c.f.	126,001		
	160,611		160,611

Profit and Loss Account

		£		£
Overhead costs				
Fuel		13,800	Gross profit b.f.	126,001
Electricity		3,870		
Water		2,965		
Admin., insurance		3,224		
Vehicle costs		5,510		
Rent, rates		6,653		
Wages		39,640		
		75,662		
Depreciation				
Plant & machinery	5,710			
Vehicles	3,254			
Buildings, glass	14,000	22,964		
Bank charges and interest		5,870		
		104,496		
Net profit		21,505		
		126,001		126,001

Balance Sheet as at 30 September 19XX

	£	£		£	£
Fixed assets			*Current liabilities*		
Plant and machinery	22,840		Creditors	1,876	
Vehicles	13,016		Bank	5,957	7,833
Buildings, glass	109,400	145,256			
Current assets			*Long term liabilities*		
Stock	36,245		Loan		37,300
Cash	797		Capital at start	122,034	
Debtors	6,374	43,416	Profit for year	21,505	143,539
		188,672			188,672

Leaffe Brothers are in partnership, both working in the business and drawing equal salaries. They also employ part-time assistants. The brothers, B. Leaffe and R.E. Leaffe, both have capital in the business but B. Leaffe, the elder brother, has the greater amount. Profit is divided equally at year end. The capital account below shows the apportionment of capital at the start and end of the year.

Capital Account for Year End 30 September 19XX

	£	£
B. Leaffe		
Capital at (start of year)	73,220	
Profit for year	10,753	83,973
R.E. Leaffe		
Capital at (start of year)	48,814	
Profit for year	10,752	59,566
Partnership capital		143,539

If drawings were made by the brothers, the amounts would be appropriately deducted from capital.

VAT

Value added tax was deliberately omitted from the exercise in order to concentrate on the main accounting principles. However, we may now consider a suitable method of coping with this ever-present problem – and the importance of simplicity is again emphasised.

Recording for VAT in the example of accounts given in this chapter would be made in:

a) the purchase account
b) the sales account
c) a personal ledger created for Customs and Excise – VAT.
d) personal ledgers

The purchase and sales accounts would each include a VAT column, similar to cash analysis, and amounts would be totalled monthly and carried to the personal ledger for Customs and Excise – VAT.

e.g. Total VAT from sales £911
Total VAT from purchases £336

Customs & Excise – VAT

Dr		£	Cr		£
26 Sept.	Bank	432	Bal. b.f.	1 Sept.	432
30 Sept.	VAT on purch.	336	30 Sept.	VAT from sales	911
		768			
30 Sept.	Cr bal. c.f.	575			
		1,343			1,343

The amount of VAT shown to be owing to Customs and Excise at the beginning of the month was paid by cheque, £432 on 26 September, and so those amounts balance out. VAT from sales is in excess of VAT on purchases, showing that £575 is owed to Customs and Excise at the end of the month.

DISCOUNT

Discount has been briefly considered at an earlier stage. Again, it has been omitted from the work in this chapter in order to maintain simplicity and convey understanding of basic principles. It is likely that discount will be given and received in a business of this type and so a suitable method of including detail is considered here. Since purchases and sales are entered in the appropriate account on receipt or production of invoice, there would be a discrepancy in the trial balance if bank payments and receipts differed from purchase and sales amounts.

Recording of discount given by suppliers and allowed on sales is by means of a discount received and discount given account. Entries are made in the same way as for a bank account. Follow the example on page 162.

Posting of discount to the correct place is difficult to follow in terms of double entry. For simplicity, the discount is entered on the same side as the bank amount. In the two examples given: Horticultural Supplies have been paid by Leaffe Brothers, the bank amount and the discount are shown as debits in the Horticultural Supplies Account, and credits in the Bank and Discount accounts; Sunny Garden Centre have paid their bill and are credited with cheque amount and discount, Leaffe Brothers' bank account would be debited as would also the discount account. The earlier chapter on double entry showed an alternative method of recording discount.

Horticultural Supplies Ltd

Dr		£	Cr		£
28 Sept.	Bank	800	Bal. b.f.	1 Sept.	860
	Discount received	60	15 Sept.	Boiler fuel	1,200
Cr bal. c.f.		1,200			
		2,060			2,060

Sunny Garden Centre

Dr		£	Cr		£
Bal. b.f.	1 Sept.	2,120	20 Sept.	Bank	2,000
3 Sept.	Shrubs	500		Dis. allowed	120
14 Sept.	Bedding plants	1,400			
			Dr bal. c.f.		1,900
		4,020			4,020

Discount received

Dr		£	Cr		£
			28 Sept.	Hortic. Supplies	60

Discount given

Dr		£	Cr		£
20 Sept.	Sunny Garden Centre	120			

Discount in the trading account

Since purchases appear in the nominal ledger and consequently in the trading account at the full invoiced price, any discount received must be shown as offsetting that cost. Discount received is a credit entry and therefore appears with sales on the credit side of the trading account. Likewise, sales appear at full value and discount given, a debit entry offsets this by appearing on the purchases, or debit, side of the trading account.

CHAPTER 18

Will the computer do it all for us?

Firstly, it must be emphasised that this brief consideration of computers and computerised accounts is not intended to be any more than just that – a brief consideration. Development of computer hardware (the actual machines) is currently so rapid that any text book on the subject is quickly out-dated. Software (computer systems and programs) development also continues apace in improving and refining, taking advantage of the continually increasing speed of computing and increasing data storage capacity. There is no attempt here to comment on the current 'state of the art'.

Will the computer do it all for us? Computer technology is readily available to do a great deal of the donkey work of accounting, but there is still a considerable amount of disciplined human effort required in even the most sophisticated systems. For the smaller business which uses a manual (hand-written) book-keeping and accounting system, the question has to be asked, 'what exactly is required?'

Computers are made for computing, and if there is not much computing to do then a manual system may be preferable. A small business with a low volume of purchases and sales, where the only requirement is an orderly set of books for presentation to the accountant at the year end, will usually be best served by a manual system (i.e. handwritten). It sometimes takes longer to prepare and enter data into a computer than to simply write the entries in a cash book or double-entry system. The computer offers the advantage that all calculations are done instantaneously and without further mental effort on the part of the operator – but if all this amounts to is totalling a few analysis columns, then the task may more easily be done with the aid of a calculator (preferably with a printer).

'What exactly is required?' was the question asked earlier. What information is required from the accounts? Is it detail and status of debtor and creditor accounts? Is it monitoring actual cash flow against budgeted cash flow? Is it allocating cost and output, to make internal transfers and to produce gross margins? Clearly, the more that is required of the computer, the more worthwhile it becomes to make the change from a manual system.

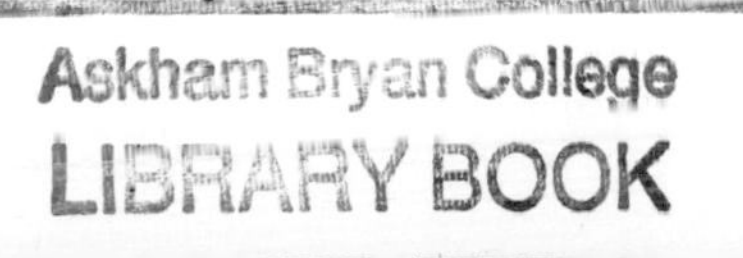

It is relatively easy and not too tedious a task to maintain a simple book-keeping system by hand, but regular calculation of such further information as outlined above may be more of a problem and consequently left for a 'more suitable moment'. Sheer volume of entries, with the resultant increased possibility of arithmetical error, presents a case for use of a computer. Correctly entered data will then be computed and automatically cross checked – an area of work which can be time consuming if errors are made in handwritten work.

The discipline imposed by using a computer is in itself a valuable move towards producing complete, balanced and legible financial data, since all too often problems occurring in a manual system are not dealt with promptly and then rear their ugly heads at year end.

Personal aptitude and attitude must be considered in cases when there is only marginal benefit to be gained from 'going computerised'. Some individuals simply do not take to the technology and a requirement to maintain their business books on computer simply adds a daunting duty to their lives. Others thoroughly enjoy and are even fascinated by computers and computing; data entry in this case may even be regarded as an enjoyable relaxation. Introduction to keyboarding and computing at an early age, as is now common with our schoolchildren, means that business folk of tomorrow will have no reservations. Who wants to draw lines and squiggles with a pen all day when the computer will make a much neater job of it?

Layout and presentation of data on computer print-out has in the past been a source of frustration and butt for many a joke. It was not easy trying to decipher yard upon yard of obscure print-out, often including unwanted information and lacking the required detail. What a pleasure it was to turn to a neatly written manual system with concisely presented data – and no wasted trees! Improved computer hardware, improved programs and, not least, programmers competing to present the most acceptable layouts are providing us with constantly improving options.

A fear held by most computer users is the possibility of sudden loss of data caused by a power cut or some other unexpected problem. Such problems can occur but are becoming progressively less likely with improved safeguards built into computer hardware.

Most important of all considerations is availability of a suitable program. If, having decided exactly what is required from the accounts, the program chosen does not come up to expectation and is difficult to use, then frustration will occur – in which case go back to square one!

There is no straightforward answer for the smaller business with low volume of transactions – personal considerations, apart from cost, do play a part here. For the business with a higher volume of transactions, then the advantages of additional selected information on demand, together with accurate work and improved print layouts, mean that sooner or later the move to use of computers for accounting is inevitable.

CHOICE OF PROGRAM

Choice of program from the many options available must be determined by the needs of the business. All too often in the past, uninformed but enthusiastic purchasers have landed themselves with a white elephant, with frustration, disillusionment and wasted time as a consequence. Advertising material can be very misleading, advantages are promoted and snags ignored or hidden. Imagine purchasing a package holiday in the sunny south somewhere and on arrival at the airport you find your flight goes to Paris and you somehow have to get to Madrid to connect with another flight in order to complete your journey! Far fetched? Maybe, but some supposedly sophisticated management accounting packages have in the past included equally tedious shortcomings. Decide what you require and ruthlessly examine the options – or live to regret it!

The following questions will help in the initial stages of determining what is required.

1) Is a book-keeping program required, or a full accounting package?
2) Is it to be cash based or double entry?
3) If cash based, is there a requirement to maintain a debtor and creditor file?
4) Is it necessary to include physical information?
5) What types of accounts are required?
6) Are enterprise gross margins and efficiency factors required?
7) Is cash flow information required?

All of the above questions can be developed relative to individual needs and additional questions asked in order to build a profile of requirements. It is also important to talk to other accounting program users in order to get to know typical current problems, and avoid the problems if possible.

Having determined what is required, gather descriptive information on a range of apparently suitable programs. Do not limit the choice to specialist agricultural (or other business) programs since there are excellent packages available which can be adapted to meet specialist requirements. Examine the descriptive material and check it against the predetermined list of requirements and short-list your choices.

Demonstration of short-listed choices is essential. Have questions prepared in advance and ask as many 'what if?' questions as possible – as long as they are relevant. Determine what training and backup is included in the price. Ask for names of other users so that their opinion can be gained. Be wary of high pressure salesmanship – a good system can be sold on recommendation alone and does not require this. More important, however, is confident and thorough knowledge of the system displayed by the demonstrator, with no apologies needed.

Custom-made programs

It may be that the profile of requirements does not seem to match satisfactorily any of the programs available and in this case the option of a custom-made package can be considered. It is unlikely that programs will be produced from scratch, but rather existing options adapted to meet special needs. Initial cost will of course be greater, but if the end result is more satisfactory, it will probably be used for a longer period, so effectively reducing the difference in cost.

OTHER USEFUL COMPUTER PROGRAMS – SPREADSHEET AND DATA BASE

There are exciting programs available which allow a 'do-it-yourself' approach to computing; they are the spreadsheet and the data base. These programs have been around for some time but they continue to develop in terms of capacity and flexibility as rapidly as other software. The two above-mentioned programs often come together with a wordprocessor program and increasingly with other facilities such as graphics.

There is enormous scope for producing business records and data storage and retrieval systems to suit individual needs, whether simple or complex. Individual designs on one spreadsheet may be linked to others on the same spreadsheet to allow progressive calculations to take place. Many successful book-keeping and accounting programs have been created by individuals who have absolutely no knowledge of computer programming, and of course there is seemingly endless scope for other applications.

The spreadsheet – how does it work?

A spreadsheet is in simple terms a matrix. Imagine a piece of graph paper which has boxes large enough to type figures in and which, if the whole thing could be seen at one time, is many times larger than the computer screen. Vertical and horizontal columns are identified by letter or number, which allows any part of the matrix to be moved instantly to the screen for inspection.

Instructions may be hidden in any box on the matrix in a simple code form, requiring figures entered in specified columns to be computed – added, subtracted, multiplied etc. – according to requirements. Instructions entered into one box may then on command be instantly 'replicated' in a row of other specified boxes across or down the matrix – amazing!

A very simple application would be creation of a cash analysis sheet. Figures entered in individual columns would automatically be totalled and appear in the 'Total' boxes. A cross checking system could be built in to ensure that the sum of the analysis columns agreed with the sum of the bank, cash and contra columns (remember your cash analysis?). Further to this, monthly totals could be moved automatically to a summary sheet, debtor and creditor adjustments carried out and final figures moved to or replicated in, for example, a profit and loss account. It would take a little patience and planning to achieve that but it would not be difficult – far more complex applications are possible.

Cash flow

Perhaps the most dramatically useful, and the most common use, of spreadsheet programs is for cash flow calculation. There really is a great saving in time and effort once the cash flow design has been entered on computer. Columns are totalled and cash flow calculated and accumulated automatically. The greatest advantage is gained on revision of cash flow data. Individual figures are altered on screen and total recalculation rapidly follows allowing a fresh new version to be printed within a very short time. Compare this to the thought of having to rewrite a manually produced cash flow; a great psychological barrier has been passed and a fresh attitude to producing this essential document is gained. The actual process of creating and entering the spreadsheet design becomes progressively easier with new programs.

Data base

Imagine a box of record cards: cow breeding records perhaps. Information is entered on each card in the sections provided and all is well, it seems. When there are large numbers of cards it becomes difficult to find individual ones quickly, despite the clever devices available. Extraction of statistics such as the number of cows in their fourth lactation or other detail may take considerable time and effort in a large herd and become another piece of potentially useful information forgone.

A computer data base program allows instant recall of individual records by simply keying in a reference number. Useful statistics can be extracted by means of listings which are predetermined by the user. Sophistication of the records and the number of listings which may be extracted will depend on the program and the data storage capacity of the computer.

Specialist programs

There are numerous specialist programs available to meet many varied business requirements. In farming circles, there are many crop and livestock programs which will aid management both in physical and technical terms as well as financial. Choice of such programs should be made very carefully after pre-determining needs, by much the same approach as that used for accounting needs, but of course the needs and objectives will be different. Do not opt for the first attractive screen display.

CHAPTER 19

Financing the business

The balance sheet must inevitably be at the centre of any consideration or negotiation of finance for the business. As already appreciated, the balance sheet shows the strengths and weaknesses in the financial structure and trends over a period of time. An understanding of options and alternatives for financing the business is essential if appraisal is to be followed by gradual improvement in the structure; this is equally important to established concerns and start-up proposals.

The long, medium and short term liabilities shown in a balance sheet represent the contribution to the financing of the business made by the owner and various lenders. The assets of the business represent the investment of these various sources of capital.

It is necessary to distinguish between owner's capital and borrowed capital but it must be clearly understood that both categories are a part of the capital required to run a business. We examine each of these sources, then their employment within the business, and cost, under the balance sheet headings.

LANDLORD'S CAPITAL – TENANT'S CAPITAL

Business capital is commonly broken down to two main categories of *landlord's* and *tenant's* capital. This concept separates the capital required for ownership of property from capital employed to run the business activity. Ownership of a farm or a shop is not a pre-requisite to business activity or farming.

Landlord's capital is invested in land and buildings. Return on capital is not normally expected to be high but increase in capital value or growth may be expected in the long term. Tenant's capital is invested in the assets which are required actually to run the business, the fixed assets and the current assets, and it is also required to finance the production process. The capital required for current assets and for the day to day operation of the business is known as **working capital.**

A further means of alternative financing is the leasing of such requirements as machinery and equipment. A lease agreement does not appear on the balance sheet as a liability and leased equipment does not appear as an asset, but the equipment is available for use without commitment of owner's or borrowed capital.

Balance Sheet – Liabilities

Long Term	*Source*	*Employment*
Loans	Private	Land
Mortgages	AMC	Land improvement
	Banks	Buildings
Medium Term		
Loans	Private	Buildings
	Banks	Machinery
	HP companies	Equipment
Short Term		
Overdraft	Banks	Working capital
Creditors	Merchants	Working capital
	Auctioneers	
	Dealers	
	Suppliers	

COST OF FINANCING – BORROWED CAPITAL

Interest is the charge paid for the privilege of borrowing money; but beware, for the manner in which interest rates are quoted can be misleading. It is essential that the basis for calculation of interest and the repayment method is clearly understood so that accurate comparisons between different sources may be made.

Bank overdraft

Interest charges on bank overdrafts are negotiable, dependent on the status of the borrower. A good bank customer with a sound business will be able to negotiate a lower margin over bank base rate than a weaker customer. The agreed interest rate is calculated on the daily balance of the overdraft.

Overdrafts are best suited to a situation where borrowing requirement fluctuates, with the advantage that when the borrowed amount is low then the interest charge is low, or zero if there is a positive balance.

Loans

There are four common methods of calculating interest, one of which is particularly costly and should be avoided. The methods are outlined below.

Reducing balance loan
A loan of this type is arranged so that the amount borrowed is reduced by equal instalments at regular intervals. The amount owing falls in a straight line if shown graphically. Interest is charged on the reducing balance and so the sum of repayment plus interest gradually reduces in proportion to the amount of interest due. The interest charged will include some compound interest, see note on APR.

A disadvantage of this type of loan is that it is inconvenient for budgeting purposes and that the largest amounts have to be paid in the early stages when it is least convenient.

Annuity loan
For convenience and simplicity of budgeting the repayments (including interest and principal) are a fixed amount which remains the same throughout the repayment period. Interest is charged on the progressively reducing balance. In the early stages of repayment the sum consists mainly of interest and only a small amount of loan repayment. With the progression of time the proportions change gradually and in the final stages the repayment amount exceeds the interest.

When interest rates increase dramatically it is advisable to adjust the instalments to ensure that the original repayment period is adhered to. Failure to adjust may result in a situation where instalments do not even cover interest charges and so the amount borrowed is increasing due to accumulating unpaid interest.

An important advantage of this type of loan in addition to the simplicity of budgeting a regular amount, is found in that the larger amount of interest charged in the early stages is an allowable charge against income tax, which particularly helps the new business. Repayment of principal increases gradually and a business would expect to be better able to cope with this as the business grows stronger.

It is particularly important to determine the APR (see following note) for annuity loans. Apparently attractive rates of interest may be quoted but further investigation reveals that they are charged on the original amount of the loan rather than the reducing balance. A simple calculation gives an approximate guide to the effect of this – but of course a more precise method is used to calculate interest charged.

A loan of £5,000 repayable over five years represents average borrowing of £2,500. Interest charged at 20 per cent p.a. on a reducing balance basis over five years would be (av. loan) £2,500 × 20% × 5 yrs = £2,500 approximately.
Interest charged at 15% flat rate on the original sum would be £5,000 × 15% × 5 yrs = £3,750 approximately.

Compound interest has been ignored in this empirical example.

Endowment loan
Interest is charged on the full amount of the loan for the whole repayment period with no periodic repayment of principal. An

endowment life assurance policy is taken out at the start of the loan period with a maturity date that coincides with repayment date of the loan. The policy is cashed and the loan repaid. A 'windfall' may be gained in the form of bonuses on the policy which effectively reduces the cost of borrowing.

ANNUAL PERCENTAGE RATE (APR)

The APR is the only reliable basis for comparison of interest rates charged by financial organisations. APR includes interest charged on accumulated interest, known as compound interest. Quoted interest rates are normally the rate used for calculation rather than the effective rate due to compound interest. When a flat rate of interest on the original sum is charged, the APR is dramatically different. Interest charged on a reducing balance at 20 per cent rate becomes 21.9 per cent when charged on a monthly basis. Where a flat rate is quoted on the original sum borrowed, the APR will be nearly double the quoted rate. Look for the APR; there is a legal requirement to give this but it is often shown in small print.

Opportunity cost – owner's capital

Investment of owner's capital must always be made with an eye for other opportunities. Investment in one activity means that the opportunity to invest elsewhere is lost, hence the term opportunity cost. Return on capital invested is compared with the best current alternatives, maybe gilts, building society or bank deposit account. When alternative new enterprises are being considered by means of partial budgeting, the alternative investment is considered to be an opportunity cost.

SOURCES OF FINANCE – BORROWED CAPITAL

Private loans

Loans from relatives and friends are an important, although declining, source of capital for farming and non-farming businesses. The advantage of such loans is mainly that interest rates tend to be lower than commercial rates and repayment arrangements may be more flexible. Informal arrangements and the death of the lender can lead to a sudden requirement to repay the loan and this could cause financial embarrassment. Preferably a formal agreement should be drawn up between the parties.

Banks

Short, medium and long term loans

Banks are the most important source of borrowing funds for farming and rural businesses. Competition for business between banks and other organisations has led to an increased range of loans offered in recent years.

Overdraft facility may be negotiated for a variety of purposes but generally it is for meeting peaks in working capital requirements. Interest charged will normally be at least 2 per cent above bank base rate but it may be much higher than this, dependent on the status and security of the business. A business which is in financial difficulty will, seemingly unfairly, have to pay a very high rate of interest.

Loans of various types and various repayment methods are available and may be designed to meet particular needs, whether short, medium or long term.

Merchants, auctioneers

Merchants in particular are an easily available and widely used source of credit. Competition necessitates normal terms of trading being allowed without charge. Failure to settle accounts within the time allowed, in order to gain extended credit, can be very costly indeed if discounts are lost or credit charges become payable. For example, if one extra month's credit is gained for the loss of a 5 per cent discount, the annual rate is 12 months × 5% = 60%. Clearly a bank overdraft would be a much better option, but it is by no means uncommon for costly merchant credit to be used in preference to discussing requirements with a bank manager.

Hire purchase

Hire purchase is mainly used for machinery but livestock schemes do exist. Credit is supplied by finance companies but the arrangements are usually carried out through the vendor. Repayment periods are usually from two to four years, with payment by standing order. Payments include capital and interest similar to an annuity loan. Ownership passes to the purchaser on payment of the deposit.

It is most important that interest rates are checked and compared with available alternatives; finance companies tend to charge higher rates than banks.

Machinery syndicates

A number of banks, in co-operation with the Federation of Syndicate Credit Companies Limited, offer special loan schemes to syndicates of two or more farmers (NFU members) for the joint purchase of machinery. Loans of up to 80 per cent of the purchase price can be arranged. Interest rates are usually competitive. Repayments may be spread over periods of three to seven years dependent on type of purchase.

Leasing

Leasing may be seen as a means of releasing capital for investment elsewhere or acquiring capital assets without the need to use borrowed capital. Ownership is retained by the leasing company which claims the capital allowance against income tax. Dependent on circumstances, the cost is usually less than that of hire purchase and occasionally less than bank borrowing.

Government assistance

A variety of grants financed either directly by the state or through the EC are available. Detail changes constantly, therefore current information should be sought from the relevant government organisation.

Proposal for a loan

A proposal for a loan from any source should be well prepared, with plan, budget and predicted cash flow. Such preparation enables the borrower to determine the wisdom and desirability of his action and it also enables the lenders when appropriate, to assess the level of risk which they are taking. Banks, the AMC and other sources of finance require detailed proposals supported by a balance sheet and budgets.

CHAPTER 20

The tax man cometh

Tax detail and regulations are subject to constant change by the government of the day, causing published specific information to be quickly out-dated. Further to this, tax issues other than the most routine are complex and are best left to the specialist. There is, therefore, no attempt to deal with taxation here other than in the most basic way and this only to give an outline on income tax as it applies to business and employees.

The tax year

The tax year runs from 6th April through to the 5th April in the following year, this applies to all individuals and businesses.

Income tax on profit arising from business

The way in which income is subject to taxation is set out in a schedule (or list). Income in the form of profit from farming is assessed under Schedule D, Case 1. Case 1 is applicable to various types of business, other cases under Schedule D are applicable to income arising from other activities such as professions.

Assessment of tax liability – business profit

Since profit arising from business activity cannot be determined until the end of the business year and, further to this, the business year may not coincide with the tax year, it follows that tax liability cannot be based on profit gained within that tax year.

The normal basis for assessment of tax liability within a tax year is therefore **the preceding year basis**. This means that tax liability in a given tax year is assessed on profit earned in the business year which ended in the previous tax year. Tax is payable in two parts, the first on 1st January and the second on 1st July.

Adjustment of accounts

Profit shown in the accounts will normally require adjustment before submission to the Inland Revenue. Various adjustments must be made, the most common being the addition back to profit of depreciation which is offset by the deduction of applicable capital allowances. Other adjustments will be made dependent on the type of business and current tax legislation.

Personal allowances are deducted from profit after other adjustments have been made and tax at the applicable rate is calculated on the balance remaining, the taxable income.

TAXATION OF EMPLOYEES

Employees are taxed under Schedule E on the Pay As You Earn basis, PAYE.

Everybody is allowed to earn a certain sum within a tax year on which no tax is payable, regardless of whether in business or employed. For the employee, the **tax free pay allowance** is allocated on a weekly cumulative basis throughout the year, i.e. the total annual allowance is divided by 52 and in the first week of the tax year 1/52 is deducted from pay to determine taxable pay, which is then taxed at the current rate. In subsequent weeks, the tax free allowance is accumulated and pay is accumulated, so that total tax liability for the year to date is recalculated every week or month. If no pay is earned for a few weeks for some reason, the tax free allowance continues to accumulate and when taxable pay is recalculated it will normally be found that liability for tax to date has fallen and a rebate is due.

In all cases the calculation is on a cumulative basis as described above, whether on manual systems or on computer (except when emergency code and week 1 basis apply).

The employer is responsible for deduction of tax and national insurance contributions from employees' pay and for payment of these deductions to Inland Revenue within the prescribed time.

The Board of Inland Revenue and Department of Social Security publish fully detailed guides for employers together with various training information. Stationery designed for calculation of PAYE is available free from Inland Revenue, but commercial versions may be preferred.

The Board of Inland Revenue publish current information on all aspects of taxation such as capital allowances, stock relief and corporation tax, but it is likely and recommended that the service of professionals be sought since their level of understanding and sources of information go far beyond the material published for general guidance.

Index

Italicised numbers refer to figures

Farming Titles From The Crowood Press